7th Meeting of the Spanish Neutron Scattering Association (SETN 2014)

Journal of Physics: Conference Series Volume 663

Pamplona, Spain
22 - 25 June 2014

ISBN: 978-1-5108-1735-7
ISSN: 1742-6588

Printed from e-media with permission by:

Curran Associates, Inc.
57 Morehouse Lane
Red Hook, NY 12571

CURRAN ASSOCIATES INC.
proceedings.com

Printed by Curran Associates, Inc. (2016)

For permission requests, please contact the Institute of Physics at the address below.

Institute of Physics
Dirac House, Temple Back
Bristol BS1 6BE UK

Phone: 44 1 17 929 7481
Fax: 44 1 17 920 0979

techtracking@iop.org

Additional copies of this publication are available from:

Curran Associates, Inc.
57 Morehouse Lane
Red Hook, NY 12571 USA
Phone: 845-758-0400
Fax: 845-758-2634
Email: curran@proceedings.com
Web: www.proceedings.com

Table of contents

Volume 663

7th Meeting of the Spanish Neutron Scattering Association (SETN)
22–25 June 2014, Pamplona, Spain

Accepted papers received: 11 November 2015
Published online: 27 November 2015

Preface

Papers

Journal of Physics: Conference Series **663** (2015) 011001 doi:10.1088/1742-6596/663/1/011001

PREFACE

The VII th Meeting of the Spanish Neutron Scattering Association was held on the campus of the Public University of Navarra (UPNa) in Pamplona (Spain) during 22-25 June 2014. It was the seventh edition of a series of biennial meetings that began in San Sebastian in 2002, which followed the meetings of Puerto de La Cruz (2004), Jaca (2006), Sant Feliu de Guixols (2008), Gijón (2010) and Segovia (2012). It is the largest meeting and discussion forum for Spanish scientific users of neutron scattering techniques, whatever the branch of science or technology development their research activity concerns.

Throughout these years, the Spanish community of neutron techniques has been consolidating, increasing every year both in the number of users and in the diversity of techniques and topics analyzed. In this sense, the series of biennial meetings of the Society aims to give visibility and summarize the activity taking place in this field. Ongoing with the initiative undertaken in the last two editions, some selected works shown in the conference are published in this edition of Journal of Physics: Conference Series.

The conference consisted of plenary lectures issued by relevant researchers in neutron science techniques, as well as invited lectures in which the most significant recent results achieved by Spanish scientists from fundamental science to applied technology were shown. To encourage the participation of as many research groups as possible and in particular young researchers, oral and poster presentations were also included.

The VII th SETN meeting was organized by the Physics Department of the Public University of Navarra in collaboration with the Spanish Society for Neutron Techniques (SETN, Sociedad Española de Técnicas Neutrónicas). The meeting attracted around 70 participants from all over the country and foreign researchers were also invited to the conference. We want to emphasize the excellent quality of the presentations and want to thank the support received from the sponsors (UPNa-Public University of Navarra, SETN -Sociedad Española de Técnicas Neutrónicas, ICMA- Materials Science Institute of Aragón, Pamplona City Council, ILL-Institute Laue Langevin, Consorcio ESS-Bilbao and ISIS-Science and Technology Facilities Council).

Finally, we want to take this opportunity to thank the scientific committee, the local organizing committee and the chairs of the conference sessions as well as the reviewers of the papers who helped with the revision process.

José Ignacio Pérez-Landazábal and Vicente Recarte

Conference Chairs

Journal of Physics: Conference Series **663** (2015) 011001 doi:10.1088/1742-6596/663/1/011001

Scientific Committee

Silvia Arrese-Igor Irigoyen (CFM, CSIC - UPV/EHU, Donostia), Javier Campo (ICMA-CSIC, Zaragoza), Carlos Frontera (ICMAB-CSIC, Barcelona), Victoria García Sakai (ISIS, Chilton), Cristina Gómez-Polo (UPNa, Pamplona), Miguel Ángel González (ILL, Grenoble), Pedro Gorría (Universidad Oviedo), Jon Gutiérrez Echevarría (EHU/UPV, Bilbao), J. Iñaki Pérez Landazábal (UPNa, Pamplona), Vicente Recarte (UPNa, Pamplona), Jesús Ruíz Hervías (UPM, Madrid), Vicente Sánchez-Alarcos (UPNa, Pamplona), Antonio Urbina (UPC, Cartagena)

Organizing Committee

J. Iñaki Pérez Landazábal (Co-Chair), Vicente Recarte (Co-Chair), Cristina Gómez-Polo, Silvia Larumbe Abuin, Vicente Sánchez-Alarcos

Editors of the Proceedings

J. Iñaki Pérez Landazábal, Vicente Recarte

Plenary speakers

Charles Simon (Institut Laue-Langevin, Grenoble, France), Miguel Angel Alario Franco (Universidad Complutense, Madrid, Spain), Dieter Richter (Jülich Centre for Neutron Science, Jülich, Germany), James Yeck (European Spallation Source, Lund, Sweden)

Invited speakers

Manu Barandiarán (BCMaterials & EHU/UPV), Arantxa Arbe (MFC, CSIC-UPV/EHU), José Luis Martínez (Consorcio ESS-Bilbao), Marta Castellote, IETcc-CSIC), Josep Lluis Tamarit (UPC), Diego Alba-Venero (ISIS), Elizabeth Castillo (CIC Energigune), Josu M. Igartua (EHU/UPV), Antonio Dos Santos (UPM), Alex Masalles (Museu Nacional d'Art de Catalunya), José Abad (UPCT), Claudia Mondelli (ILL), Oscar Fabelo (ILL), Aurora Nogales (IEM-CSIC), Jesús Rodríguez (UC), Gerardo Garcés (CENIM-CSIC), Elena Forcén (ICMA-CSIC), Pedro Gorria (U. Oviedo)

Journal of Physics: Conference Series **663** (2015) 011003 doi:10.1088/1742-6596/663/1/011003

Peer review statement

All papers published in this volume of *Journal of Physics: Conference Series* have been peer reviewed through processes administered by the proceedings Editors. Reviews were conducted by expert referees to the professional and scientific standards expected of a proceedings journal published by IOP Publishing.

On the exchange bias effect in NiO nanoparticles with a core(antiferromagnetic)/shell (spin glass) morphology

N. Rinaldi-Montes[1], P. Gorria[1], D. Martínez-Blanco[2], A. B. Fuertes[3], L. Fernández Barquín[4], J. Rodríguez Fernández[4], I. de Pedro[4], M. L. Fdez-Gubieda[5], J. Alonso[5], L. Olivi[6], G. Aquilanti[6], I. Puente-Orench[7], J. A. Blanco[1]

[1]Departamento de Física, Universidad de Oviedo, E-33007 Oviedo, Spain
[2]Servicios Científico-Técnicos, Universidad de Oviedo, E-33006, Oviedo, Spain
[3]Instituto Nacional del Carbón, CSIC, E-33080 Oviedo, Spain
[4]*CITIMAC*, Facultad de Ciencias, Universidad de Cantabria, E-39005, Santander, Spain
[5]Univ. Basque Country, EHU, Dept. Elect. & Elect. and BCMaterials, E-48080 Bilbao, Spain
[6]Sincrotrone Trieste S.C.p.A., S.S.14 Km 163.5, 34149 Basovizza, Trieste, Italy
[7]Instituto de Ciencia de Materiales de Aragón, CSIC-Universidad de Zaragoza and Institut Laue-Langevin, BP 156, F-38042 Grenoble Cedex 9, France

E-mail: nataliarin@gmail.com

Abstract. The unexpected appearance of magnetic hysteresis and exchange bias effects in nominally antiferromagnetic NiO nanoparticles is usually explained in terms of a core/shell morphology, where a spin glass-like shell is exchange coupled to an antiferromagnetic core. However, recent studies have challenged the validity of this assumption for small enough NiO nanoparticles. In this work we present proof of the core/shell model for NiO nanoparticles with sizes below 10 nm by combining neutron powder diffraction and magnetic measurements. In addition, we have verified that the exchange bias effect persists even when the particle size is reduced down to 4 nm.

1. Introduction

In the recent years, magnetic nanoparticles (NPs) have attracted an intense research effort because they appear to play by a different set of rules compared to those governing bulk materials [1]. As size reduction down to nanometric scale usually changes the magnetic properties of the system, NPs emerge as promising candidates in different applications such as catalysis, solar cells, supercapacitors and biomedicine [2]. In order to optimize the performance and efficiency of the NPs, a deep understanding of their physicochemical properties is needed. A feature common to all the NPs is that their magnetic behavior is strongly influenced by finite size effects. Many studies point out the importance of surface anisotropy as the main cause for the surface spins to become non-magnetic or orientated in a different direction from the core spins [3]. In particular, NPs made up of materials which are antiferromagnetic (AFM) in their bulk form constitute suitable systems for studying finite size magnetism. As AFM materials need a high degree of symmetry in order to maintain a compensated magnetic structure, their magnetic behavior experiences noticeable changes when reducing size.

Bulk NiO is a paradigmatic AFM material with a Néel temperature (T_N) of approximately 523 K. However, the anomalous appearance of a ferromagnetic (FM) moment in nominally AFM NPs was first predicted by Néel in the early 1960s [4] and later confirmed by numerous experimental works [5, 6]. The most widely accepted theory explains this feature as arising from a core/shell model, where the core spins are antiferromagnetically aligned and the outer shell spins are in a disordered spin glass (SG)-like state [6, 7]. One of the most remarkable consequences of this core/shell morphology is the occurrence of the exchange bias (EB) effect. EB emerges as a consequence of the exchange interaction at the interface between two magnetic phases with different anisotropy energies. It manifests as a shift of the magnetic hysteresis loop along the field axis when the system is cooled down in an external magnetic field through the ordering temperature of the most anisotropic phase. However, a recent study has reported that NiO NPs with an average diameter of 7 nm present a core(AFM)/shell(SG) structure without any trace of EB effects [8]. Hence, the assumption that the occurrence of the EB effect is a direct consequence of the core/shell model could be no longer appropriate for small enough NP sizes.

In this work we have performed a neutron powder diffraction and magnetometry joint investigation in order to check the correlation between the core/shell morphology and the occurrence of the EB effect in NiO NPs with sizes below 10 nm.

2. Experiment

Two samples of NiO NPs assemblies (NiO-4nm and NiO-9nm) were synthesized via the pyrolysis of a nickel nitrate taking place under N_2 within the restricted volume formed by the porosity of a commercial activated carbon. The average size of the NPs was controlled by means of the heating temperature. NiO-4nm and NiO-9nm were prepared at 300 °C and 400 °C, respectively. High-resolution transmission electron microscopy (HRTEM) images of each sample were obtained under a JEOL-JEM-2100F microscope aiming to determine the size, shape and aggregation state of the NPs. The crystalline structure of the samples was investigated by powder x-ray diffraction (XRD) using a Philips X'Pert PRO MPD diffractometer with Cu $K\alpha 1$ radiation ($\lambda = 1.5406$ Å). Room temperature powder neutron diffraction (ND) patterns were recorded at the Institut Laue-Langevin (D1B instrument) using a wavelength of $\lambda = 2.52$ Å. Magnetization data were measured on a Quantum Design PPMS-9T magnetometer.

3. Results and Discussion

Figures 1(a)-1(b) display representative HRTEM images of the samples showing the monodispersity, crystallinity and quasi-spheroidal shape of the NPs. The fit of size histograms to lognormal distributions provides values for average NP diameter and standard deviation, $D(\sigma)$, of 4(1) nm and 9(1) nm for samples NiO-4nm and NiO-9nm, respectively.

Rietveld refinement of the XRD patterns [see Figures 1(c)-1(d)] was performed using FullProf package. XRD data were fitted according to a rocksalt $Fm\bar{3}m$ crystal structure. The average NP diameters obtained for NiO-4nm and NiO-9nm coincide with those estimated from HRTEM histograms. The value of the cell parameter obtained for both samples is 4.18(1) Å, very close to that reported for bulk NiO ($a_{NiO}^{bulk} = 4.178$ Å).

ND patterns in Figures 1(e)-1(f) were modelled according to a $Fm\bar{3}m$ crystal structure and a propagation vector $\mathbf{k} = (\frac{1}{2}, \frac{1}{2}, \frac{1}{2})$ referred to a cubic cell [9]. During the fitting process, the unit cell parameters and the scale factors were constrained so as to have the same values in the nuclear and magnetic phases. The obtained values for the magnetic moment per Ni ion (μ) are $1.4(1)\mu_B$ and $1.7(1)\mu_B$ for samples NiO-4nm and NiO-9nm, respectively, thus smaller than the value reported for bulk NiO ($2\mu_B$) [9]. As the number of Ni atoms was fixed to be the same in

Journal of Physics: Conference Series **663** (2015) 012001 doi:10.1088/1742-6596/663/1/012001

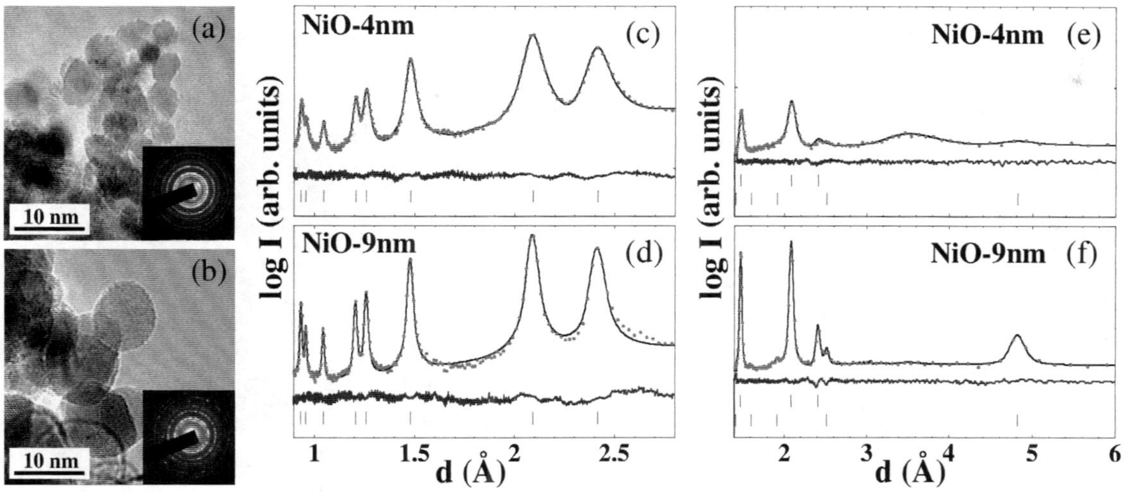

Figure 1. (Color online) (a-b) HRTEM images of samples NiO-4nm (top) and NiO-9nm (bottom). Insets display selected area electron diffraction (SAED) patterns. (c-d) XRD and (e-f) ND patterns in logarithmic vertical scale measured at room temperature. Red dots and black solid line represent the experimental data and the Rietveld fit, respectively. Blue lines correspond to the difference between experimental and calculated patterns. For ND patterns, vertical markers indicate the position of nuclear (upper row) and magnetic (lower row) Bragg reflections, while for XRD patterns only vertical markers corresponding to nuclear reflections are shown.

both the nuclear and magnetic phases, the effective reduction in the expected magnetic moment per Ni atom can be linked to the presence of a magnetically disordered shell at the surface of the NPs [10].

Figures 2(a)-2(b) show the temperature dependence of the dc magnetization measured under zero-field-cooling (ZFC) and field-cooling (FC) conditions between 2 K and 300 K. The $M_{ZFC}(T)$ and $M_{FC}(T)$ curves of both samples reveal identical features, so that three characteristic temperatures of the system can be identified: (i) a sharp peak at low temperature ($T \sim 9$ K); (ii) a broad maximum centred at $T \sim 58$ K (155 K) for NiO-4nm (NiO-9nm); and (iii) a splitting temperature between $M_{ZFC}(T)$ and $M_{FC}(T)$ curves at $T \sim 97$ K (300 K) for NiO-4nm (NiO-9nm). This phenomenology in NiO NPs can be well-explained in the frame of the core/shell model. Then, each NP would be formed by: (i) an AFM ordered core with uncompensated magnetic sublattices (giving rise to a net magnetic moment) due to the reduction in the number of exchange coupled moments caused by finite size effects; and (ii) a magnetically disordered shell, which thermally fluctuates like a SG system [6, 7].

Finally, the magnetic hysteresis loops measured at $T = 5$ K under ZFC and FC ($H_{cool} = 80$ kOe) conditions are depicted in Figures 2(c)-2(d). The shift of the loops toward both the negative field (the EB field, H_{EB}, increases from 0.90 kOe for NiO-4nm to 2.75 kOe for NiO-9nm) and the positive magnetization axis, together with an enhanced coercive field constitute strong evidence of the existence of an EB effect [see Figures 2(e)-2(f)]. The EB origin can be understood starting out from the core(AFM)/shell(SG) morphology derived from ND and dc magnetization measurements. Therefore, the exchange coupling at the interface between the AFM core and the SG shell arises as the most likely mechanism for the occurrence of the EB effect.

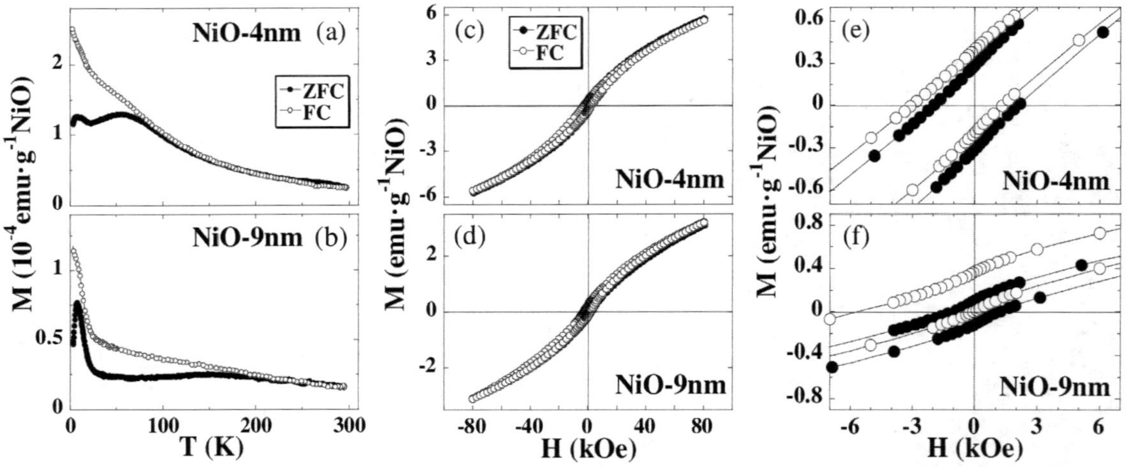

Figure 2. (a-b) Temperature dependence of the magnetization measured under ZFC and FC conditions for samples NiO-4 nm (top) and NiO-9nm (bottom) under an applied magnetic field of 100 Oe. (c-d) Magnetic hysteresis loops measured at $T = 5$ K after cooling the samples under ZFC and FC ($H_{cool} = 80$ kOe) conditions. (e-f) Enlarged view of the central part of the hysteresis loops evidencing the characteristic features of the exchange bias effect.

4. Conclusions

We have shown that the exchange bias effect in NiO nanoparticles with average diameters below 10 nm arises as a consequence of the core(antiferromagnetic)/shell(spin glass) morphology. The magnitude of the exchange bias effect decreases when reducing particle size due to the smaller core-to-shell volume ratio, but still persists for diameter values as small as 4 nm.

Acknowledgments

Financial support has been received from Spanish MINECO (Project MAT2011-27573-C04) and MECD (Grant FPU12/03381).

References

[1] Dormann J L and Fiorani D 1992 *Magnetic Properties of Fine Particles* Edited by North Holland, Amsterdam.
[2] Teja A S and Koh P Y 2009 *Prog. Cryst. Growth Charact. Mater.* **55** 22
[3] Iglesias O and Labarta A 2001 *Phys. Rev. B* **63** 184416
[4] Néel 1949 *Physica* **15** 225
[5] Kodama R H, Makhlouf S A and Berkowitz A E 1997 *Phys. Rev. Lett.* **79** 1393
[6] Winkler E, Zysler R D, Mansilla M V, Fiorani D, Rinaldi D, Vasilakaki M and Trohidou K N 2008 *Nanotechnology* **19** 185702
[7] Rinaldi-Montes N, Gorria P, Martinez-Blanco D, Fuertes A B, Barquin L F, Fernandez J R, de Pedro I, Gubieda M L F, Alonso J, Olivi L, Aquilanti G and Blanco J A 2014 *Nanoscale* **6** 457
[8] Cooper J F K, Ionescu A, Langford R M, Ziebeck K R A, Barnes C H W, Gruar R, Tighe C, Darr J A, Thanh N T K and Ouladdiaf B 2013 *J. Appl. Phys.* **114** 083906
[9] Roth W 1958 *Phys. Rev.* **110** 1333
[10] Martinez B, Obradors X, Balcells L, Rouanet A and Monty C 1998 *Phys. Rev. Lett.* **80** 181

A Monte Carlo simulation code applied to diffraction experiments with polarization analysis

L A Rodríguez Palomino[1,2], A Stunault[1], J Dawidowski[2], L Temleitner[3], L Pusztai[3], G J Cuello[1]

[1] Institut Laue Langevin, 71, av des Martyrs, 38042 Grenoble, France
[2] Consejo Nacional de Investigaciones Científicas y Técnicas, Centro Atómico Bariloche and Instituto Balseiro, Bustillo 9500 (R8402AGP) Bariloche, Argentina
[3] Institute for Solid State Physics and Optics, Wigner Research Centre for Physics, Hungarian Academy of Sciences, H-1121 Budapest, Konkoly Thege út 29-33., Hungary

E-mail: cuello@ill.eu

Abstract.
Neutron diffraction with polarization analysis is a very useful technique to separate the coherent and incoherent contributions to the scattering cross section. This experimental technique allows the direct measurement of the spin-flip and non spin-flip contributions, which in turn are related to the coherent and incoherent contributions by simple linear combinations. Nevertheless, the correction procedures to obtain the single coherent elastic scattering are tricky. We present a method, based on Monte Carlo simulation, that allows one performing these corrections in a wide variety of coherent and incoherent samples. In this work, we have applied this method to the study of heavy water, for which we obtained the interference structure factor. We describe the experimental setup to perform the measurements in a reactor source and we detail the experimental protocol allowing the determination of the non spin-flip and spin-flip diffractograms.

1. Introduction

The use of neutron diffraction techniques on hydrogenated materials is limited because of the huge incoherent scattering cross section of hydrogen (^{1}H or H). For this reason, diffraction experiments on this kind of systems are performed on deuterated samples, where H is substituted with deuterium (^{2}H or D). This substitution reduces the incoherent contribution, but it has two drawbacks. First, one looses the unique information provided by a nucleus with a negative scattering length, for example, negative peaks in the radial distribution functions. Second, if isotopic effects are expected, one can never be sure whether a given feature observed in the correlation function is coming from an isotopic effect or is indeed a real feature of the correlation function.

The best way to overcome this problem is directly measure the incoherent contribution over a relevant momentum transfer. In particular, for the case of liquid systems, this relevant momentum transfer range should extend beyond 15 Å^{-1} in order to obtain relatively good Fourier transforms of the structure factors [1]. The access to this momentum transfer range is quite limited because only few instruments worldwide have enough hot neutron flux. But how can one measure the incoherent scattering directly? The answer is that such measurement is almost impossible, at least if the aim is the determination of the total incoherent contribution. Nevertheless, a fraction of the incoherent part, the spin-incoherent scattering due to the disorder of the nuclear spins, can be determined using polarized neutrons. While coherent scattering does not change the neutron spin (non-spin-flip scattering), spin-incoherent has the neutron spin reversed (spin-flip scattering). In the case of H, over 99% of the incoherent signal is due to spin-incoherent scattering. However, this technique has been little exploited so far in studies of liquids, as the Q-range allowed by the existing instruments was too small [2, 3].

Now we present a study were we have used polarized hot neutrons in order to extract the spin-incoherence on a mostly coherent liquid: the heavy water. In such a system, the incoherent contribution is small and, consequently, the corrections are small too, and the results more accurate than in a system with more incoherent scattering. This rather simple case is used to assess the method, before extending it to highly incoherent (hydrogenated) systems. This ultimate evolution is still under development and will be treated in a forthcoming publication.

Our method is based on a Monte Carlo (MC) code simulating the interaction of the neutrons with the atoms, which allows us to follow the history of the individual neutron until its detection. This method is very useful in order to evaluate the corrections (multiple scattering, attenuation, inelasticity) as a whole and not like independent corrections as usual [4]. The drawback is that it requires a model of the interactions between the neutrons and the nuclei of the particular system giving the probability of scattering at a given angle and a given energy exchange. In the case of water, we use the Synthetic Model proposed by Granada [5], which has very well shown its usefulness in the past [6].

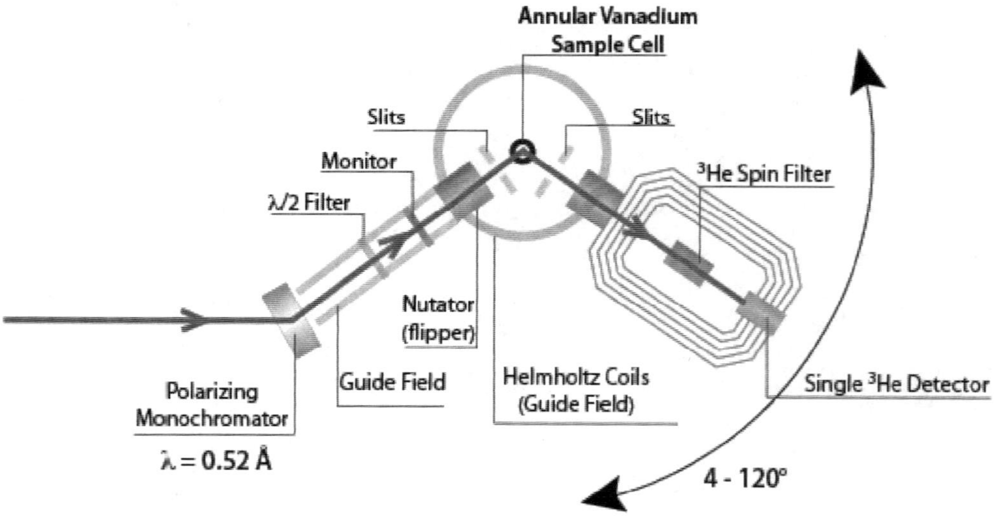

Figure 1. Sketch of the D3 setup.

2. Experimental

The experiment was performed on the Spin Polarized Hot Neutron Beam Facility (D3) at the ILL (Institut Laue Langevin, Grenoble, France) [7]. This diffractometer is the only one worldwide allowing polarization analysis at short-wavelengths and covering a wide range of momentum transfer Q (0.8 to 21 Å^{-1} at a neutron wavelength of 0.52 Å). Figure 1 shows a sketch of the D3 setup with its different components. The polarizing monochromator is a Heusler crystal, Cu_2MnAl ((111) Bragg reflection). Between the polarizer and the sample position, a guide field preserves the spin state of the neutron and a monitor allows normalization of the eventually detected intensities to the actual neutron flux. Before and after the sample nutators and slits respectively reverses (flips) the spin state of the neutron (whether necessary) and adjust the beam size. The neutrons scattered by the sample are analyzed using a polarized ^{3}He spin filter [8, 9], which uses the drastically different transmissions of the neutrons with a spin state parallel or antiparallel to the spin state of the helium atoms to serve as a polarization analyzer. The transmitted neutrons then reach the single ^{3}He detector. The detector arm rotates around the sample, covering a total angular range of 4° to 120°. In our experiments, we chose steps of 0.5° up to 26° and 1° for the higher angles and typical data acquisition times of 1 minute/step, half the time spent on measuring the non-spin-flip (NSF), spin-flip (SF) neutron intensities respectively. Finally a beam stop is placed in the direction of the direct beam from the monochromator, to stop the transmitted neutrons and minimize the radiation background.

The measurement was performed using a sample of D_2O in liquid state at room temperature, provided by the ILL chemistry lab. The sample was loaded into a cylindrical double walled vanadium cell of 10.7 mm external diameter, 0.15 mm walls thickness and 58 mm height; thus, the sample thickness was 1.2 mm, corresponding to a volume of 0.829 cm^3 in the beam. The mass was checked before and after the experiment, confirming that the container had no leakage.

Measurements of NSF and SF diffractograms were carried out for 24 hours. Shorter times were employed to measure the instrument background, the signal from the empty vanadium sample cell and to calibrate the polarization. All diffractograms were normalized to the monitor counts. The decay of the polarization (and efficiency) of the ^{3}He spin filter was monitored every few hours by measuring the SF and NSF intensities from the (111) reflection of a cylindrical germanium crystal, mounted on top of the vanadium sample cell and automatically driven into the beam via a vertical translation. The measured polarization efficiencies, between 0.65 and 0.75 with a relaxation time of the order of 80 to 90 hours, were used to correct the data for the time dependent analyzer efficiency. Finally all SF and NSF diffractograms were corrected for the absorption by the empty filter.

3. Data treatment

The first step in data processing consists in determining the experimental NSF and SF intensities ($I^{S}_{\rm exp}(Q)^{\rm NSF,SF}$) scattered by the sample, which can be determined in a first approximation as the difference between the sample+container minus the empty container intensities. Figure 2 shows these magnitudes as obtained from our experiment on D_2O. It is worth noting that the SF empty cell intensity is comparable with the SF total intensity, making the background subtraction a delicate issue.

Our MC code was initially developed to correct standard (non-polarized) neutron diffraction data [12] and has now been extended to include the special case of polarized neutrons. Equations (25) and (26) in Ref. [12] are extended to the case of a polarized incident beam and an unpolarized sample, thus the angular distributions of scattered neutrons (SF and NSF) after the first scattering are given by the following expressions

$$I_1(\mathbf{k_0},\theta)^{\rm NSF,SF} = \frac{[1-t(E_0)]\ \Sigma_{\rm s}(\mathbf{k_0})^{\rm NSF,SF}}{\Sigma_{\rm t}(\mathbf{k_0})}\ \frac{\sigma(E_0,\theta)^{\rm NSF,SF}}{\sigma_{\rm s}(E_0)^{\rm NSF,SF}}, \tag{1}$$

$$\tilde{I}_1(\mathbf{k_0},\theta)^{\rm NSF,SF} = \frac{V}{A(\hat{\mathbf{k}}_0)}\ \Sigma_{\rm s}(\mathbf{k_0})^{\rm NSF,SF} \int {\rm d}E\ \varepsilon(E)\ H_1(\mathbf{k_0},\mathbf{k})\ \frac{\sigma(E_0,E,\theta)^{\rm NSF,SF}}{\sigma_{\rm s}(E_0)^{\rm NSF,SF}}, \tag{2}$$

where $I_1(\mathbf{k_0},\theta)$ represents the angular distribution of single scattering without attenuation, which is equivalent to a point-like sample with the same scattering power as the whole sample, and $\tilde{I}_1(\mathbf{k_0},\theta)$ represents the angular distribution of neutrons detected after the first scattering. In these expressions, $t(E_0)$ is the transmission of the neutron beam at energy E_0, $\Sigma_{\rm t}(\mathbf{k_0})$ is the macroscopic total cross section, $\Sigma_{\rm s}(\mathbf{k_0})^{\rm NSF,SF}$ is the macroscopic scattering cross section,

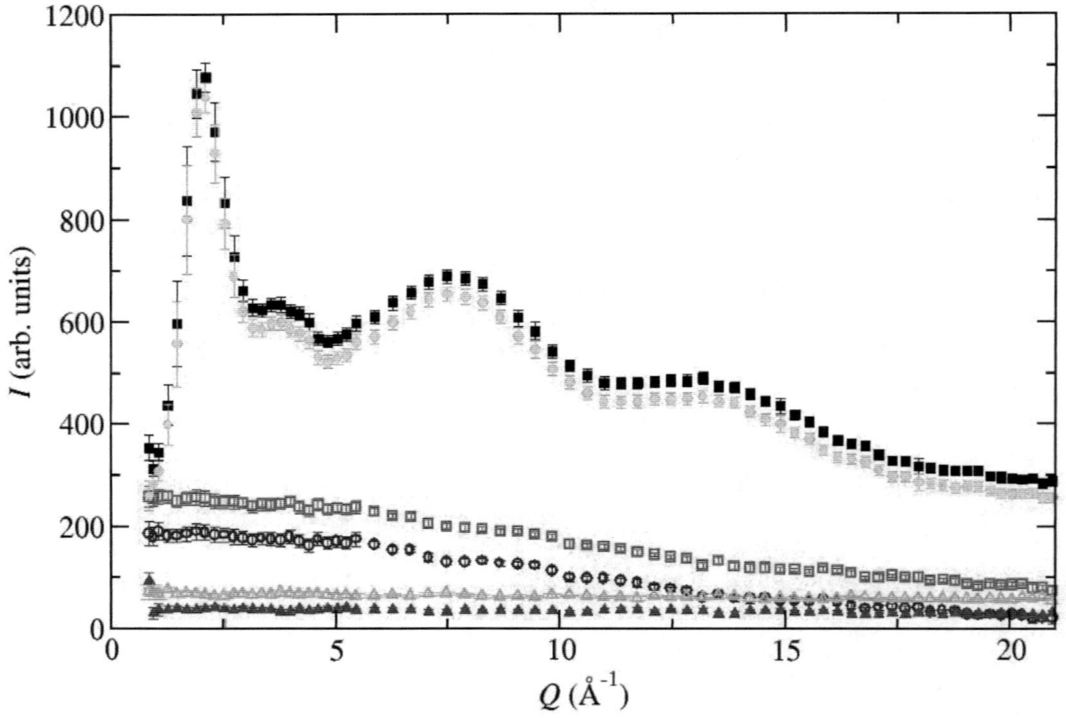

Figure 2. Experimental intensities from the D_2O sample (circles), sample+container (squares) and vanadium container (triangles). The open and full symbols represent the SF and NSF contributions, respectively.

$\frac{\sigma(E_0,\theta)^{\mathrm{NSF,SF}}}{\sigma_s(E_0)^{\mathrm{NSF,SF}}}$ is the normalized microscopic differential cross section, V the sample volume, $A(\hat{\mathbf{k}}_0)$ is the cross sectional area exposed to the incident beam, $\frac{\sigma(E_0,E,\theta)^{\mathrm{NSF,SF}}}{\sigma_s(E_0)^{\mathrm{NSF,SF}}}$ the normalized microscopic double-differential cross section, $\varepsilon(E)$ in the detector efficiency at energy E, and $H_1(\mathbf{k}_0, \mathbf{k})$ is the primary attenuation factor defined by Sears [13]. This factor takes into account the attenuation of the incident neutron beam before the first scattering and then on its way towards the detector.

These magnitudes, Eqs. (1) and (2), are not accessible by the experiment, but they can easily be simulated. Using the neutron transport theory, Sears [13] determined the analytical expressions for the multiple scattering. These expressions, involving multiple integrals, make impractical their numerical quantification, which is why we perform this evaluation through MC numerical simulation.

Finally, the corrected data are normalized to absolute units (barn/sterad/atom) using the measured differential cross section from the vanadium sample container as a reference, and taking into account the difference in volume and density of the sample and vanadium, i.e., the different number of scatterers in the beam. The normalization constant is chosen in order to obtain the good high-Q limiting value for the vanadium cross section, that is, the free scattering cross section (4.83 barns).

4. Monte Carlo simulation

The MC simulation allows the assessment of the different components of the scattering distributions (single scattering, multiple scattering, attenuation factors and the total scattering), which in turn will serve to determine a Q-dependent correction factor. This factor is then employed to correct the NSF and SF experimental intensity, leading to their corresponding single scattering experimental intensity, as described by Eq. (1) and (2).

The MC simulation is based on the method proposed by Bischoff [14] and Copley [15], which describes all the interaction processes between the neutron and the sample since entering in the sample until it is registered by the detector at a given angle. Discrete neutron histories are tracked and averaged in a random walk directed by interaction probabilities obtained either by measurements or by models.

As input, the MC program requires information on the sample and the experimental setup. For the sample, the following information is required: the macroscopic total cross section, the absorption probabilities as a function of energy, the scattering laws, the sample geometry, and the SF interaction probability [17], which in the case of D_2O is $p^{\mathrm{SF}} = \sigma_{\mathrm{s}}^{\mathrm{SF}}/\sigma_{\mathrm{s}} = 0.1395$ [16]. For the instrument, the incident neutron energy and the beam dimensions are needed.

4.1. Neutron histories

The neutron trajectory starts at a randomly chosen point in the sample and the first collision is also randomly determined taking into account the path length distribution function. At each collision the new energy and flight direction are determined using probability distributions. For the energy exchange we usually use Granada's Synthetic Model [5], which allows the estimation of the normalized double-differential cross section. In this work we employ a hybrid model which works very well for a variety of coherent and incoherent samples. This model has been previously tested to perform the data treatment in standard neutron diffraction experiments [4, 12] and now has been extended for the case of diffraction experiments using polarized neutrons. For a given collision the flight direction is determined using the experimental angular distribution, whereas the energy after the collision is determined using Granada's model.

4.2. Neutron weight

The implemented MC program is based on an implicit simulation, which is more efficient and reduces the computing time as compared to analogue simulations. In the implicit simulation the neutron never gets out the sample being forced to scatter (since absorption is forbidden in this altered scheme) [14]. In this scheme, to compensate the biases in the probability distribution employed, a weight is assigned to each neutron, which decreases with the transmitted fraction along the path, the initial value being 1. The neutron history is finished when the weight drops under a predetermined cut-off value and after applying the Russian Roulette criterion [14]. For the NSF and SF scattering processes, the neutron weight of the collision i can be calculated from the weight of the preceding collision $i-1$ as follows

$$w_i^{\mathrm{NSF,SF}} = w_{i-1}^{\mathrm{NSF,SF}} \, [1 - t(E,d)] \, \frac{\Sigma_{\mathrm{s}}(E,0)^{\mathrm{NSF,SF}}}{\Sigma_{\mathrm{t}}(E,0)}, \tag{3}$$

where $1-t(E,d)$ represents the probability of interaction within the distance d, and $\frac{\Sigma_{\mathrm{s}}(E,0)}{\Sigma_{\mathrm{t}}(E,0)}$ gives the probability that the neutron is not absorbed when it experiences a NSF or SF scattering process.

4.3. Scoring

For each collision process, the contribution of the current history to the final diffractogram is scored by the detector placed at an angle θ with respect to the incident beam. In the single scattering case, the contribution of a point r to the (NSF, SF) neutron distribution after the first scattering can be calculated by scoring the estimator

$$I_1(\mathbf{r}_1, \mathbf{k}_0, \theta)^{\mathrm{NSF,SF}} = w_1 \; D(Q)^{\mathrm{NSF,SF}}, \tag{4}$$

where $D(Q)$ is the normalized experimental angular distribution expressed in the elastic-Q scale ($Q_\mathrm{e} = 2k_0 \sin(\theta/2)$). Similarly (NSF, SF) neutron distribution detected after the i-th scattering can be calculated by scoring the estimator

$$\tilde{I}_i(\mathbf{r}_i, \mathbf{k}_{i-1}, \theta)^{\mathrm{NSF,SF}} = w_i \; t(E_{i-1}, \mathbf{r}_i, -\hat{\mathbf{k}}) \; \varepsilon(E) \; D(Q)^{\mathrm{NSF,SF}}, \tag{5}$$

where $t(E_{i-1}, \mathbf{r}_i, -\hat{\mathbf{k}})$ is the transmission coefficient from the point $\mathbf{r}_i$ inside the sample to the sample surface in the direction $\hat{\mathbf{k}}$ to the detector position. It is noteworthy that for the special case $i = 1$, the expressions (4) and (5) averaged over a large number of histories converges to the expressions (1) and (2). Using this formalism, the multiple scattering contribution can then be calculated as:

$$\tilde{I}_\mathrm{M}(\mathbf{r_1}, \mathbf{k_0}, \theta)^{\mathrm{NSF,SF}} = \sum_{i \geq 2} \tilde{I}_i(\mathbf{r}_i, \mathbf{k_0}, \theta)^{\mathrm{NSF,SF}}. \tag{6}$$

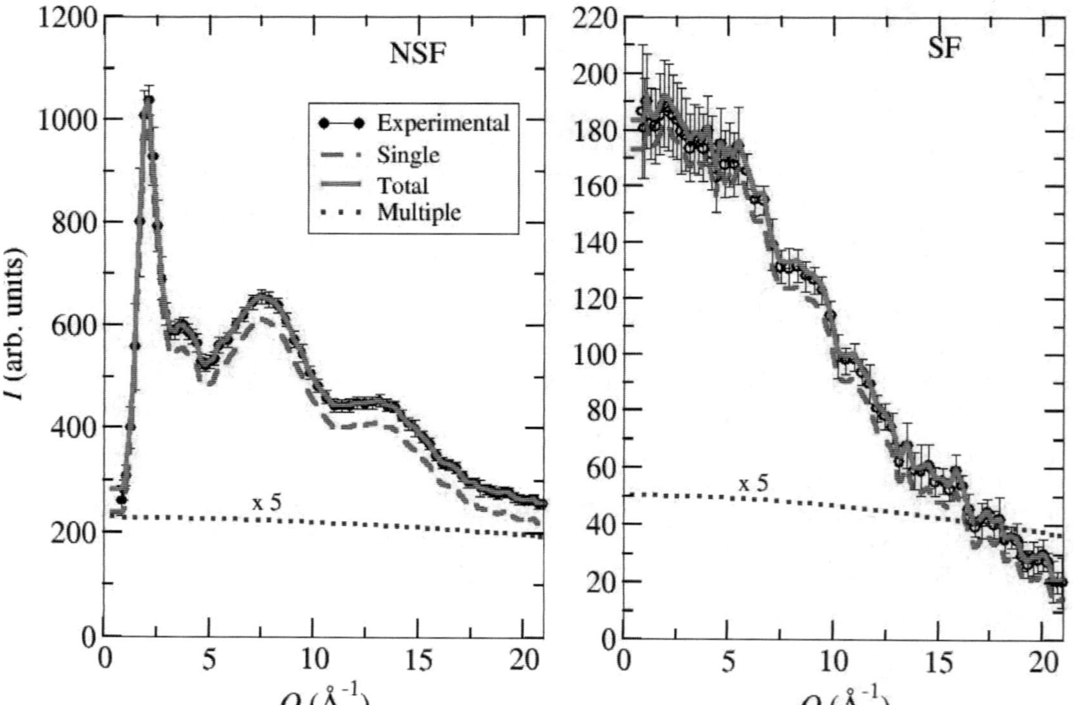

Figure 3. Experimental (NSF/SF) intensities (symbols) compared with the Monte Carlo simulation (solid line). The single (dashed line) and multiple (dotted line) scattering are also shown; the latter has been multiplied by a factor 5 for the sake of visibility.

Journal of Physics: Conference Series **663** (2015) 012002 doi:10.1088/1742-6596/663/1/012002

4.4. Correction factor

Once the different contributions are obtained by means of the numerical simulations, the NSF and SF correction factors can be calculated as

$$F(\mathbf{k_0}, \theta)^{\text{NSF,SF}} = \frac{\tilde{I}_1(\mathbf{k_0}, \theta)^{\text{NSF,SF}}}{\tilde{I}_1(\mathbf{k_0}, \theta)^{\text{NSF,SF}} + \tilde{I}_{\text{M}}(\mathbf{k_0}, \theta)^{\text{NSF,SF}}} \frac{1}{A(\mathbf{k_0}, \theta)}, \tag{7}$$

where $A(\mathbf{k_0}, \theta) = \frac{\tilde{I}_1(\mathbf{k_0},\theta)^{\text{NSF,SF}}}{I_1(\mathbf{k_0},\theta)^{\text{NSF,SF}}}$ is the attenuation factor. This correction factor is then used to correct the experimental intensities, thereby obtaining the microscopic differential cross section according to the expression (1), which is the sought magnitude in all diffraction experiments.

4.5. Iterative procedure

The correction process is based on an iterative method, where the outputs in each run are employed to calculate the new experimental angular distributions, in its turn serving as new input in the next run until no variation in the multiple scattering components is observed. In this case, with heavy water as the sample, the convergence is achieved after 3 iterations and each run consists in 200000 neutron histories. The iterative scheme implemented to correct the experimental distributions is

$$D_i(Q)^{\text{NSF,SF}} = D_0(Q)^{\text{NSF,SF}} \ F_{i-1}(\mathbf{k_0}, \theta)^{\text{NSF,SF}}, \tag{8}$$

where the subscript i indicates the iteration number, and $D_0(Q)^{\text{NSF,SF}}$ is the measured experimental distribution.

5. Results and discussion

Figure 3 shows the NSF (left) and SF (right) experimental intensities for D_2O and its different scattering components as obtained in the last iteration (single, multiple and total scattering). We found a good agreement between the experimental data and simulations and we also observe that almost all the measured signal is due to the single scattering ($\sim$ 95 %). This is explained by the fact that the mean free path of the incident neutron (0.302 eV) is about 27 mm which is much larger than the total sample thickness (2.6 mm). The oscillations observed in the SF signal (right panel in Fig. 3) have a statistical origin due to the low incoherence of D_2O.

Once the NSF/SF correction factors are determined, the experimental intensities corrected by multiple scattering and attenuation can be expressed as

$$I_{\text{exp}}^{\text{corr}}(Q)^{\text{NSF,SF}} = I_{\text{exp}}^{\text{S}}(Q)^{\text{NSF,SF}} \ F(\mathbf{k_0}, \theta)^{\text{NSF,SF}}. \tag{9}$$

A simple linear combination of the corrected NSF and SF intensities then gives the corrected coherent $I_{\text{coh}}^{\text{corr}}(Q)$ and incoherent $I_{\text{inc}}^{\text{corr}}(Q)$ experimental intensities:

$$I_{\text{coh}}^{\text{corr}}(Q) = I_{\text{exp}}^{\text{corr}}(Q)^{\text{NSF}} - \frac{1}{2} I_{\text{exp}}^{\text{corr}}(Q)^{\text{SF}}, \tag{10}$$

$$I_{\text{inc}}^{\text{corr}}(Q) = \frac{3}{2} I_{\text{exp}}^{\text{corr}}(Q)^{\text{SF}}. \tag{11}$$

The left panel of Fig. 4 shows the corrected NSF and SF experimental intensities, while the right panel shows the coherent and incoherent experimental intensities.

The expression (10), obtained after the experimental corrections and normalization process, is still affected by coherent inelastic effects and isotope incoherent scattering (the latter is also

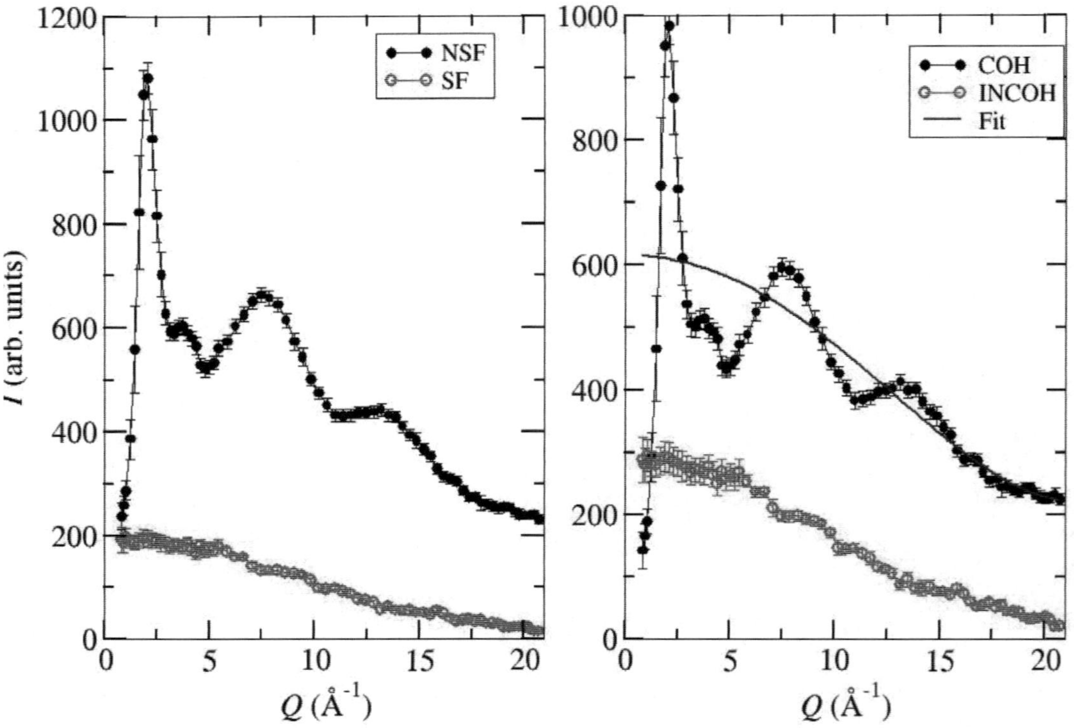

Figure 4. The corrected NSF/SF experimental intensities (left) and the coherent/incoherent experimental intensities (right). The Placzek polynomial fitted to the coherent contribution (Eq. (12)) is also shown.

NSF and is not eliminated by the method). In order to correct for these effects, we use a Placzek polynomial [18], as follows:

$$p(Q) = 615.06 - 1.32\, Q^2 - 0.037\, Q^3 + 0.0027\, Q^4. \tag{12}$$

Finally, subtracting this polynomial (Fig. 4, right panel) from the total coherent contribution (Eq. (10)), the normalized coherent structure factor $F_N(Q)$ is obtained. In Fig. 5 we compare our coherent structure factor with that obtained by Soper [19]. We observe a good overall agreement, specially in what concerns the intensity in absolute units and the peak positions, except for the low-Q region and the peak close to 4 Å^{-1}. These small discrepancies are likely related to problems in the background subtraction, having more impact on the low-angle region of the diffractogram and which needs to be improved. However, they do not prevent the demonstration of the usefulness of the method.

6. Conclusions

We presented here a consistent method to correct data from polarized neutron diffraction experiments, which can be employed in a wide variety of coherent and incoherent samples. The proposed method corrects for multiple scattering, attenuation effects and inelasticity in a consistent way via the MC simulation of neutron trajectories, and this for both NSF and SF diffractograms. At present this method is applied to the case of D_2O where the incoherent contribution is low, but it is currently being extended to highly incoherent samples. The results

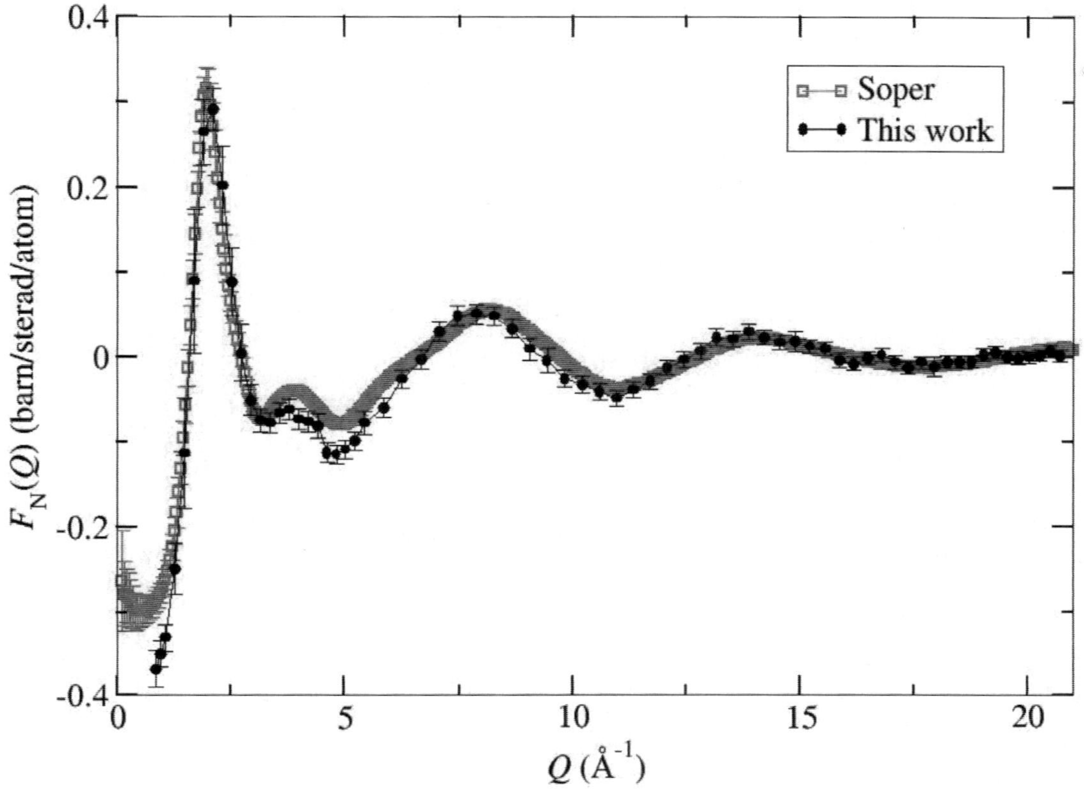

Figure 5. The structure factor of D_2O after applying our correction method. Our results are compared with those obtained by Soper [19].

we obtained for D_2O are comparable with those obtained by Soper [19], showing that our procedure produces consistent data.

Regarding discrepancies observed in Fig. 5 at low-angle region, a modification of the MC code is foreseen in order to take into account the attenuation of the container intensity at small angles, which will improve the outcome of our method.

Concerning the instrument D3, we show its usefulness in the precise determination of the incoherent contribution, which in turn conducts to a better determination of the interference structure factor. The main drawback is the limited counting rate for an instrument that was designed for diffraction experiments on single crystal. In order to improve the registered intensity and to reduce the counting time, a possible solution would be the use of a larger detector covering a wider range of scattering angles, as is the case of D4 [20] at ILL. Increasing the solid angle covered by the detector would increase the statistics and thus reduce the error bars.

It is important to note that a Placzek correction is still necessary, because we can not take into account the inelastic coherent scattering. The main improvement of this method is the ability of measuring a big fraction of the incoherent scattering, which makes the Placzek correction smaller, producing more accurate structure factors.

At the core of the MC method there is the knowledge of a model describing the energy exchanges between the neutron and the nuclei. We used Granada's Synthetic Model [5] for water, but it could be extended for other hydrogenated molecular liquids.

We hope that this work, and others currently in preparation or submitted [21], will help showing that the use of polarized neutrons is extremely useful for the study of hydrogenated liquids. This technique is particularly interesting for biological systems, where the isotopic substitution of hydrogen with deuterium is not always possible or desirable.

Acknowledgements

The authors thank Sebastien Vial for the technical support during the experiments on D3. The beam time provided by the Institut Laue Langevin (Grenoble, France) is gratefully acknowledged (proposal 6-02-519). LT and LP wish to thank the Hungarian Basic Research Fund (OTKA) for financial support via grant No. K083529.

References

[1] G. J. Cuello, J. Phys.: Cond. Matter **20**, 244109 (2008).
[2] J. C. Dore, J. H. Clarke, J. T. Wenzel, Nucl. Instr. Meth. **138**, 317 (1976).
[3] L. Temleitner, L. Pusztai, W. Schweika, J. Phys.: Cond. Matter **19**, 335207 (2007).
[4] J. Dawidowski, G. J. Cuello, L. A. Rodríguez Palomino, Neutron Diffraction ISBN 978-953-51-0307-3, InTech (2012)
[5] J. R. Granada, Phys. Rev. B **31**, 4167 (1985).
[6] J. R. Granada, V. H. Gillette, R. E. Mayer, Phys. Rev. A **30:12**, 5594 (1987).
[7] A. Stunault, D3 web page at the Institut Laue Langevin, Grenoble, France, `http://www.ill.eu/d3` (2015).
[8] J. Dreyer, L. P. Regnault, E. Bourgeat-Lami, E. Lelièvre-Berna, S. Pujol, F. Thomas, M. Thomas, F. Tasset, Nucl. Instr. Meth. Phys. Res. A **449**, 638 (2000).
[9] W. Heil, J. Dreyer, D. Hofmann, H. Humbol, E. Lelièvre-Berna, F. Tasset, Phys. B **267-268**, 328 (1999).
[10] H. H. Paalman, C. J. Pings, J. Appl. Phys. **33**, 2635 (1962).
[11] I. A. Blech, B. L. Averbach, Phys. Rev. **137**, A1113 (1965).
[12] L. A. Rodríguez Palomino, J. Dawidowski, J. J. Blostein, G. J. Cuello, Nucl. Instr. Meth. Phys. Res. B **258**, 453 (2007).
[13] V. F. Sears, Adv. Phys. **24**, 1 (1975).
[14] F. G. Bischoff, PhD thesis, Rensselaer Polytechnic Institute (1969).
[15] J. R. D. Copley, P. Verkerk, A. A. van Well, H. Fredrikze, Computer Phys. Comm. **40**, 337 (1986).
[16] J. Dawidowski, J. R. Granada, J. R. Santisteban, F. Cantargi, L. A. Rodríguez Palomino, Experimental Methods in the Physical Sciences **Neutron Scattering-Fundamentals** ISBN 978-0-12-398374-9 **44**, Elsevier Inc (2013)
[17] G. L. Squires, Introduction to the Theory of Thermal Neutron Scattering ISBN 0-486-69447-X, Dover Publications Inc, Mineola, New York (1978).
[18] G. Placzek, Phys. Rev. **86**, 377 (1952).
[19] A. K. Soper, Hindawi Article ID 279463, 67 (2013).
[20] H. E. Fischer, G. J. Cuello, P. Palleau, D. Feltin, A. C. Barnes, Y. S. Badyal, J. M. Simonson Appl. Phys. A: Material Sci. Processing **74**, S160 (2002).
[21] L. Temleitner, A. Stunault, G. J. Cuello, L. Pusztai, Phys. Rev. B **92**, 014201 (2014).

Journal of Physics: Conference Series **663** (2015) 012003 doi:10.1088/1742-6596/663/1/012003

Morin transition in Hematite nanoparticles analyzed by neutron diffraction

J. I. Pérez-Landazábal[1*], C. Gómez-Polo[1], V. Recarte[1], S. Larumbe[1], V. Sánchez-Alarcos[1], M. Fernandes Silva[2], E.A Gómez Pineda[2], A.A. Winkler Hechenleitner[2], M.K. Lima[2] and J.A. Rodriguez-Velamazán[3,4]

[1]Departamento de Física, Universidad Pública de Navarra, Campus de Arrosadía 31006 Pamplona, Spain.

[2]Departamento de Química, Universidade Estadual de Maringá, 87020-900 Maringá-PR, Brazil

[3]Instituto de Ciencia de Materiales de Aragón, CSIC - Universidad de Zaragoza, Zaragoza, Spain

[4]Institut Laue-Langevin, CRG's D1B, F-38042 Grenoble, France

*E-mail: ipzlanda@unavarra.es

Abstract. Hematite (α-Fe_2O_3) undergoes a first order spin reorientation transition called the Morin transition: upon cooling, the moments align antiferromagnetically along the rhombohedral axis, and the net magnetic moment goes to zero. Morin transition temperature is around to 260K in bulk materials and depends on the mean particle size. In this work, the Morin transition has been studied by neutron diffraction as function of temperature and applied magnetic field in 47 nm nanoparticles. The Rietveld analysis of the diffraction spectra around the Morin transition shows a similar behavior to that found in bulk samples. On the other side, the magnetic field induced phase transformation has been analyzed.

1. Introduction

Because of their inherent environment-friendly feature, iron oxide materials have been extensively studied in diverse fields including catalysis [1,2], environment protection [3], gas sensors [4,5] magnetic storage media [6], and clinical diagnosis and treatment [7]. In order to pursue an excellent performance, many efforts have also been carried out on the synthesis and properties of iron oxide nanostructures [1,4,8,9]. Iron oxide is an important transition metal oxide which usually exists in nature in three phases: magnetite (Fe_3O_4), maghemite (γ-Fe_2O_3) and hematite (α-Fe_2O_3). Hematite is the most stable form of iron oxide (α-Fe_2O_3). Its electronic, catalytic and magnetic properties are of increasing interest from the application point of view in different fields [10]. At room temperature, hematite is weakly ferromagnetic with a rhombohedral corundum structure (Space group R-3c) [11]. The magnetic moments are approximately perpendicular to the rhombohedral axis with a small canting of the magnetic sublattices, giving a small net magnetic moment. Upon cooling, the material undergoes a first order spin reorientation transition, the Morin transition. The moments align antiferromagnetically along the rhombohedral axis and the net magnetic moment goes to zero. Morin transition temperature T_m, is around to 260 K in bulk materials [12,13]. However, T_m depends on the mean particle size, d. When the size is smaller than 100 nm, T_m decreases abruptly [14,15]. This effect is due to the increase of the surface-volume ratio of the particles, which yields to a lack of symmetry at

Journal of Physics: Conference Series **663** (2015) 012003 doi:10.1088/1742-6596/663/1/012003

nanometric scale. Therefore, particle size reduction hinders the magnetic interaction producing the long-range antiferromagnetic order. For nanoparticles smaller than 10 nm, the antiferromagnetic order does completely vanish [16]. In this work, the Morin transition has been studied by neutron diffraction as function of temperature and applied magnetic field. The Rietveld analysis of the diffraction spectra around the Morin transition shows a similar behavior to that found in bulk samples. On the other side, the magnetic field induced phase transformation has been analyzed "In situ" in the diffractometer.

2. Experimental

Nanoparticles of hematite were elaborated by modified aqueous sol-gel method [17]. Saturated aqueous metal nitrate solution was prepared and maintained at room temperature under stirring for two hours. After that, it is heated under vigorous stirring until total water evaporation. The temperature was maintained at 250 °C until the total thermal degradation. The nanostructured material was obtained after burning the precursor powder under air atmosphere at 400 °C. The mean diameter size is 47 nm determined by Scherrer formula after instrumental correction but particles show irregular shapes. Neutron diffraction experiments were performed on the D1B (CRG-D1B-12-249). The Rietveld Method as implemented in the the Fullprof [18] program has been used to analyze the neutron spectra.

3. Results and Discussions

Figure 1 shows a three-dimensional picture (intensity, angle 2θ and temperature) of the phase evolution observed "in situ" at the diffractometer. At room temperature, the diffraction pattern corresponds to hematite (Space group R-3c). The transition to the antiferromagnetic structure can be detected at around T_m=240K through the pure magnetic (003) reflection measured at 32° (rhombohedral setting). The peak intensity linked to the ferromagnetic phase lowers as the temperature decreases. Results show the absence of changes in the transformation temperature when compared with bulk alloys. On the other side, a multidomain structure is expected to be present due to the particle size and then interparticle interaction could be neglected [19].

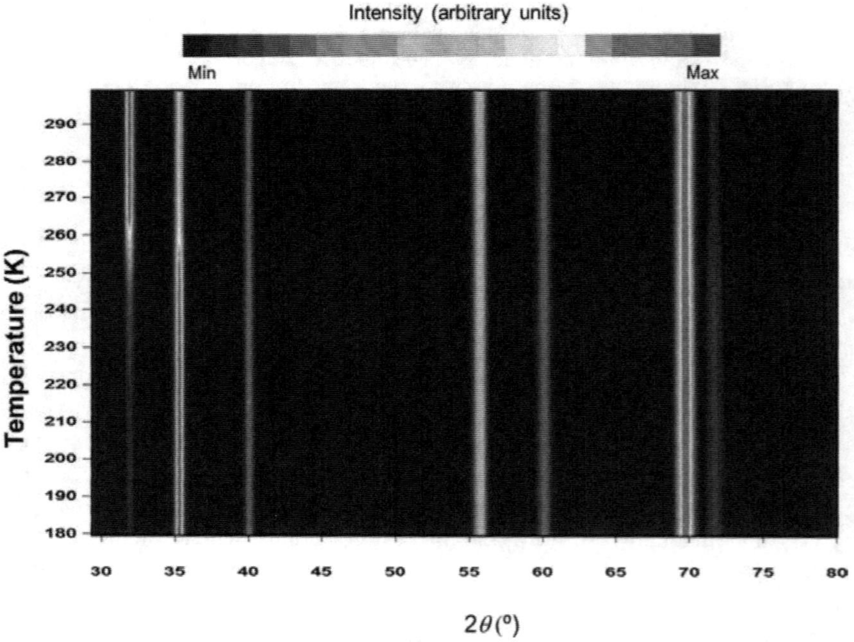

Figure 1: Thermo-diffraction patterns in a mesh-plot corresponding to the hematite nanoparticles of 47 nm size. The colour code indicates different intensities (in arbitrary units) from blue (minimum) to red (maximum). The Morin transition is clearly noticed by an increase of the intensity of the reflection at 2θ *ca.* 32° and a decrease of the reflection at 2θ *ca.* 35°.

Figure 2 shows the observed and calculated neutron powder diffraction patterns from D1B for α-Fe_2O_3 at 280 K (above) and 200 K (below). The difference between observed and calculated is shown below the diffraction patterns, with tick marks indicating the Bragg reflection positions from the nuclear and magnetic contribution. The (003) Bragg reflection is also indicated.

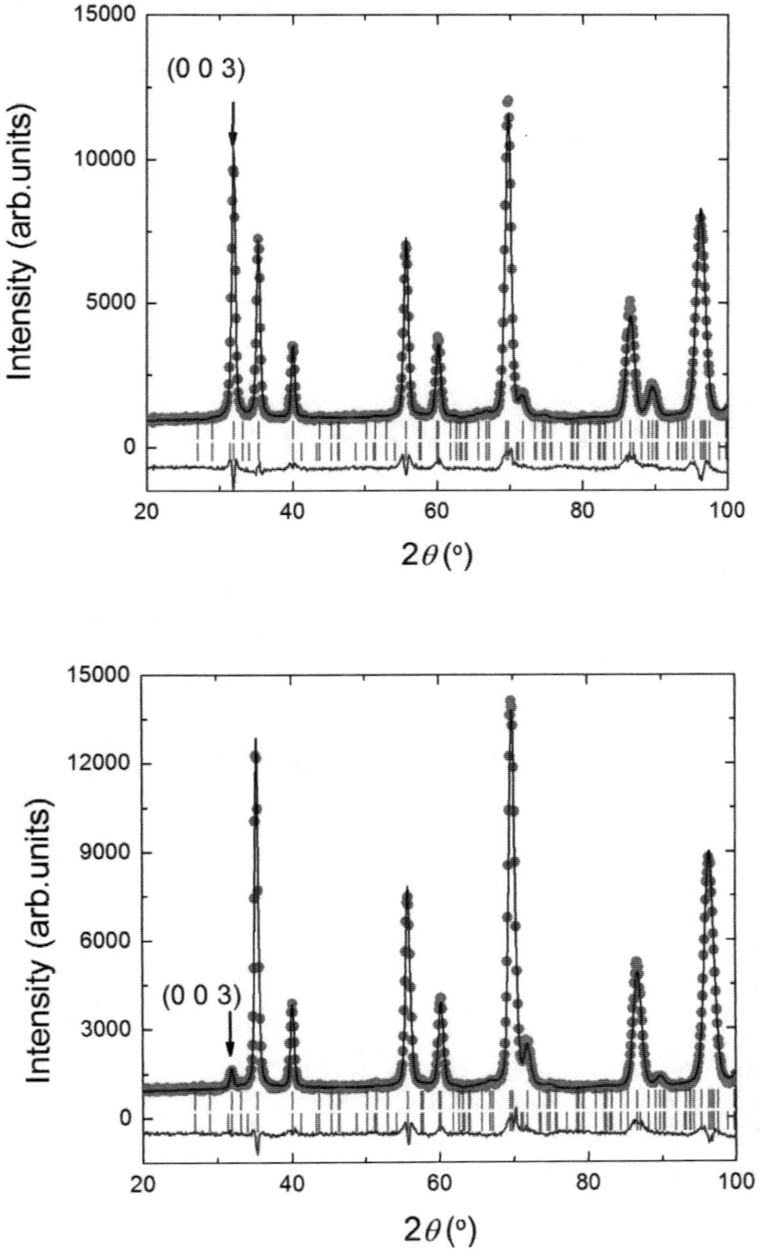

Figure 2: Refined patterns for T = 280 K ($T>T_M$)-Above and T = 200 K ($T<T_M$)-Below. The experimental data have been represented as red circles, the calculated curve as solid black line and the difference between them as solid blue line. The green vertical lines represent the Bragg positions for the nuclear (top) and magnetic (bottom) phases.

Journal of Physics: Conference Series **663** (2015) 012003 doi:10.1088/1742-6596/663/1/012003

Crystallographic and magnetic information obtained at 10 K and 280 K are shown in Table 1. As the symmetry of the system does not allow to determine the direction of the magnetic moment in the hexagonal basal plane, we can only distinguish between the in-plane and out-plane components of the magnetic moment. Above T_m, the magnetic moments are approximately perpendicular to the rhombohedral axis with a small canting of the magnetic sublattices with the scattering plane (around 17°), giving a small net magnetic moment. Upon cooling below 250 K, the moments align antiferromagnetically along the rhombohedral axis and the net magnetic moment goes to nearly zero. This change is mainly reflected in the the intensity of the (003) peak changing radically to a much smaller value below T_m, (see figure 1) although at least six reflections have important magnetic contribution reducing the magnetic Bragg error.

	10 K	270 K
a (Å)	5.0254(1)	5.0298(1)
c (Å)	13.7215(6)	13.7372(8)
Magnetic moment on plane (μB)	0.869(35)	3.566(24)
Magnetic moment out of plane (μB)	3.840(22)	1.115(76)
R_{mag}(%)	7.91	6.62

Table 1: Refined cell parameters and in-plane and out of plane magnetic moments at 10 K (below T_m) and 270 K (above T_m).

Nevertheless, the moments are not perfectly aligned with the rhombohedral axis. The Rietveld refinement of the structure at low temperature gives an estimate of the angle at which the moments tilt out away from the [111] direction. This angle is around 12.6 ° at 10 K, which agrees with previously published values [20,21].

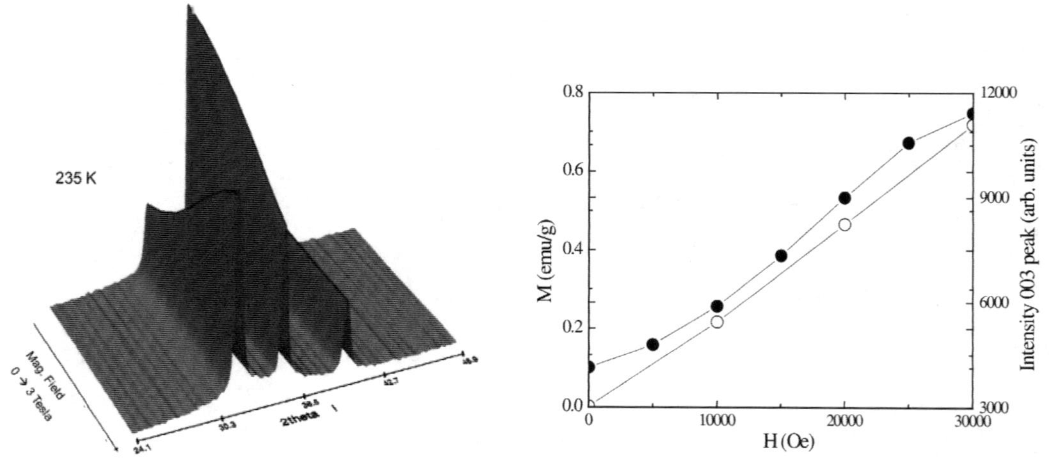

Figure 3: Magnetic field induced Morin transition recorded at 235 K (Left) and magnetization [22] and the (003) reflection intensity as a function of the applied magnetic field (Right).

Finally, regarding the magnetostructural transition, an increase in magnetization upon heating across the Morin transition has been observed previously [22]. The magnetization change during the transformation has been estimated as ΔM=0.3 emu/g for the maximum applied field of H=60 kOe. Besides, a shift of the Morin transition to lower temperatures as the applied magnetic field increases, indicates that the magnetic field stabilizes the high temperature phase with respect to the low temperature one. Thus, the Morin transition should be induced by an external magnetic field in a particular temperature range. Figure 3 (left) shows, for example, the effect of a magnetic field application at T= 235 K. The sample is in the antiferromagnetic state at zero field but the application of a 30 kOe leads to the weak ferromagnetic phase as shown by the large increase of the (003) reflection intensity. This direct observation evidences the magnetic field induced transition from anti to ferromagnetic behavior. Figure 3 (Right) shows a comparison between the magnetization measured by squid magnetometry [22] and the (003) reflection intensity as a function of the applied magnetic field.

4. Conclusions

Neutron diffraction experiments performed "*In Situ*" allow determining the magnetic transition that occurs in Hematite nanoparticles. The Rietveld analysis of the diffraction spectra around the Morin transition shows a similar behavior to that found in bulk samples. On the other side, the magnetic field induced phase transformation from anti to ferromagnetic state has been observed.

Acknowledgements

The Institute Laue-Langevin and SpINS are acknowledged for the allocated neutron beamtime (CRG-D1B-12-249). JARV acknowledges CSIC for a JAEdoc contract.

References

[1] Tsodikov M V, Rostovshchikova T N, Smimov V V, Kiseleva O I, Maksimov Y V, Suzdalev I P and Ikorskii V N 2005 *Catal. Today* **105** 634.
[2] Zheng Y H, Cheng Y, Wang Y S, Bao F, Zhou L H, Wei X F, Zhang Y Y and Zheng Q 2006 J. *Phys. Chem. B* **110** 3093.
[3] Zhong L S, Hu J S, Liang H P, Cao A M, Song and Wan L J 2006 *Adv. Mater.* **18** 2426.
[4] Yang Y,Ma H X,Zhuang J, and Wang X 2011*Inorg. Chem.* **50** 10143.
[5] Chen J, Xu L N, Li W Y and Gou X L 2005 *Adv. Mater.* **17** 582.
[6] Zeng H, Li J, Liu J P, Wang Z L and Sun S H 2002 *Nature* **420** 520.
[7] Gupta A K and Gupta M 2005 *Biomaterials* **26** 3995.
[8] Bronstein L M, Huang X L, Retrum J, Schmucker A, Pink M, Stein B D and Dragnea B 2007 *Chem. Mater.* **19** 3624.
[9] Herrera F, Lopez A, Mascolo G and Albers E, Kiwi J 2001 *Appl. Catal. B* **29** 147.
[10] Navrotsky A, Mazeina L and Majzlan J 2008 *Science* **319** 1635.
[11] Gupta R K, Ghosh K, Dong L and Kahol P K 2010*Mater. Lett.* **64** 2132.
[12] Bengoa J F, Alvarez A M, Bianchi A E, Punte G, Vandenberghe R E, Mercader R C and Marchetti S G 2010 *Mater. Chem. Phys.* **123** 191
[13] Zysler R D, Fioran Di. Testa A M, Suber L, Agostinelli E and Godinho M 2003*Phys. Rev. B* **68** 212408.
[14] Özdemir O, Dunlop D J, and Berquó T S 2008 *Geochemistry Geophysics Geosystems* **9** Q10Z01.
[15] Amin N and Arajs S 1985*Phys. Rev. B* **35** 4810.
[16] Bødker F and Mørup S 2000*Europhys. Lett.* **52** 217 .
[17] Fernandes D M, Winkler Hechenleitner A A, Lima S M, Andrade L H C, Caires A R L and Gómez Pineda E A 2011Mater. *Chem.Phys.* **128** 371.
[18] Rodriguez-Carvajal J 1993 *Physica B* **192** 55.

[19] Frandsen C, Lefmann K, Lebech B, Bahl C R H, Brok E, Ancoña S N, Theil Kuhn L, Keller L, Kasama T, Gontard L C and Morup S 2011 *Phys. Rev. B* **84** 214435

[20] Krén E, Molnàr B, Svàb E, Zsoldos E 1974 *Solid State Commun.* **15** 1707.

[21] Hill A H, Jiao F, Bruce P G, Harrison A, Kockelmann W and Ritter C 2008 *Chem. Mater.* **20** 4891.

[22] Pastor J M, Pérez-Landazábal J I, Gómez-Polo C, Recarte V, Larumbe S, Santamarta R, Fernandes Silva M, Gómez Pineda E A, Winkler Hechenleitner A A and Lima K M 2012 *Appl. Phys. Lett.* **100** 063102.

Journal of Physics: Conference Series **663** (2015) 012004 doi:10.1088/1742-6596/663/1/012004

Structure determination of Fe-Al-Ge alloys

D Gargicevich[1,2], V M Galván Josa[2,3], C Blanco[2,3], O A Lambri[1,2], G J Cuello[4]

[1] Laboratorio de Materiales, Escuela de Ingeniería Eléctrica, Centro de Tecnología e Investigación Eléctrica, Facultad de Ciencias Exactas, Ingeniería y Agrimensura, UNR, Av. Pellegrini 250, 2000 Rosario, Argentina.
[2] Consejo Nacional de Investigaciones Científicas y Técnicas (CONICET),
[3] Universidad Nacional de Córdoba, Argentina.
[4] Institut Laue Langevin, 71, Av. des Martyrs, 38042 Grenoble, France.

E-mail: `cuello@ill.eu`

Abstract. We studied the crystalline structure of Fe - 8at.%Al - 4at.%Ge alloy between 300 and 1300 K and its relation to the mechanical response by means of neutron diffraction and mechanical spectroscopy. At room temperature we observe a Fe_3Al-type ordered structure with a deficiency of Al in the 8*c* sites. The Ge atoms are distributed in the 4*a* and Al atoms in 8*c* sites. At high temperature we observe an order-disorder transformation when the crystal structure becomes Fe-α type. This loss of order gives rise to the hysteresis behavior of damping between the heating and cooling runs.

1. Introduction

The use of Fe-Al-Ge alloys in technological applications depends mainly on its brittle and creep behavior at high temperatures. It is of special interest to know the evolution of the structural phases with temperature in this ternary system, since it has a strong influence on dislocations and grain boundaries mobility [1, 2, 3]. The information about the structure of ternary Fe-Al-Ge system is very scarce, being it related to a partial isotherm [4], but the lack of data concerning its mechanical and magnetic properties is even deeper.

In this work, we combine neutron thermo-diffraction and mechanical spectroscopy (MS, damping and modulus measurements as a function of temperature), including amplitude-dependent damping analysis, in order to reveal the phase transitions and their effects on the mechanical properties of the Fe - 8at.%Al - 4at.%Ge alloy. The amplitude-dependent damping is a term that encompasses the damping processes in systems for which this damping is dependent on the deformation. In our analysis we study a magnitude (S) measuring the strength of these particular damping processes.

2. Experimental

We used polycrystalline samples of Fe - 8at.%Al - 4at.%Ge, which were homogenized at 1323 K during 1 h under high vacuum (10^{-2} Pa), and then quenched in room-temperature water. Then, we cut the samples in small parallelepiped bars (around $0.3 \times 0.3 \times 1.5$ mm^3) using a cooled kerosene low-speed saw. We give more details about the sample preparation elsewhere [5].

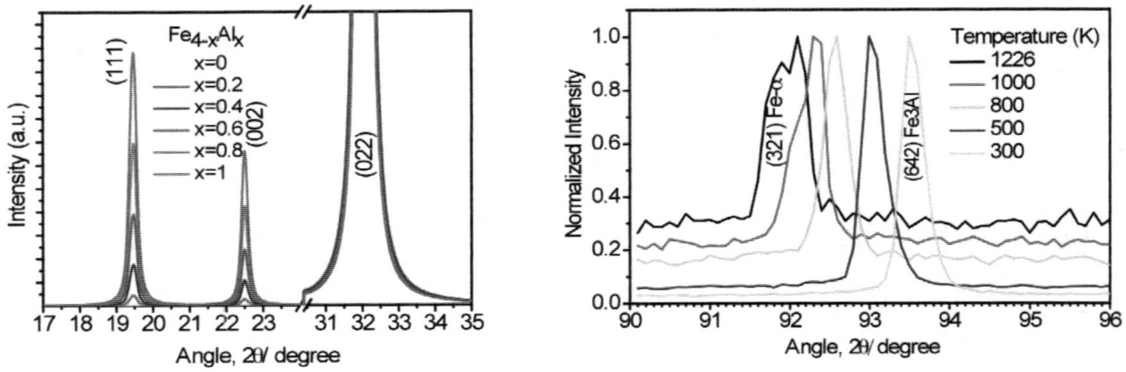

Figure 1. (Left panel) Theoretical (111) and (002) reflections of the ordered $Fe_{4-x}Al_x$ structure as a function of occupation factor of Al in the $8c$ sites. (Right panel) Measured patterns around 92° for Fe - 8at.%Al - 4at.%Ge alloy. At temperatures around 1000 K, the reflection (321) of the disordered Fe-α phase type is clearly distinguished.

We performed neutron powder thermo-diffraction (NPD) experiments at the high-flux D20 diffractometer [6] at the Institut Laue-Langevin, using high-intensity mode and a wavelength of 1.12 Å. We used vanadium cylinders as sample containers, which were placed in a standard vanadium furnace (a cylindrical heating element of 35 mm in diameter). This furnace was in turn located in a vacuum chamber (10^{-2} Pa) and samples heated with a ramp of 3 K/min. During this heating process we registered a diffractogram every 5 min, then integrating 15 K. Thanks to the large banana-like detector, each diffractogram covers a scattering angle range of about 150°. We analyzed the diffraction patterns using standard software, such as FULLPROF [7] for Rietveld refinement and Topas® for texture effects with spherical harmonics approximation.

We performed MS measurements at frequencies close to 1 Hz, under high vacuum (about 10^{-5} Pa), with free decaying vibrations during subsequent heating and cooling runs at 1 K/min [8, 9, 10, 11, 12]. We used samples with a parallelepipedal shape ($0.5 \times 2 \times 20$ mm^3) and with a maximum strain of $\epsilon_0 = 5 \times 10^{-5}$. We fitted the area of free decaying vibrations in logarithmic scale with polynomials minimizing the χ^2. The derivative of the decaying areas (in log scale) as a function of period number [8, 13, 14, 15] gives the damping Q^{-1} as a function of ϵ_0. We obtain the strength of the amplitude-dependent damping effects (S) as $S = \Delta Q^{-1}/\Delta\epsilon_0$, where Q^{-1} is the damping change measured when the amplitude changes $\Delta\epsilon_0$. In fact, S is an average slope value of Q^{-1} against ϵ_0 curves, in the whole strain interval; a positive value indicates that the greater is the deformation , the greater is the damping.

3. Results

Since the studied samples consist in a few small parallelepipeds, *i.e.* they are not isotropic powders, there exists a low number of domains satisfying the diffraction condition at each scattering angle. This has two consequences: (1) the presence of a strong texture effect, and (2) changes in the intensities of the diffracted peaks due to small sample movements associated with dilatations during heating. In order to reduce these effects we taken the average of eight consecutive patterns for each temperature range.

The Fig. 1 (left panel) shows simulated diffractograms for $Fe_{4-x}Al_x$ compounds (space group $Fm\bar{3}m$) with the aluminum atom at $8c$ sites. It is apparent that the intensities associated with the Bragg planes with high-Al content decrease with the Al occupation factor. If the occupation factor of Al is less than 20%, the (111) reflection almost disappears. Since the aluminum

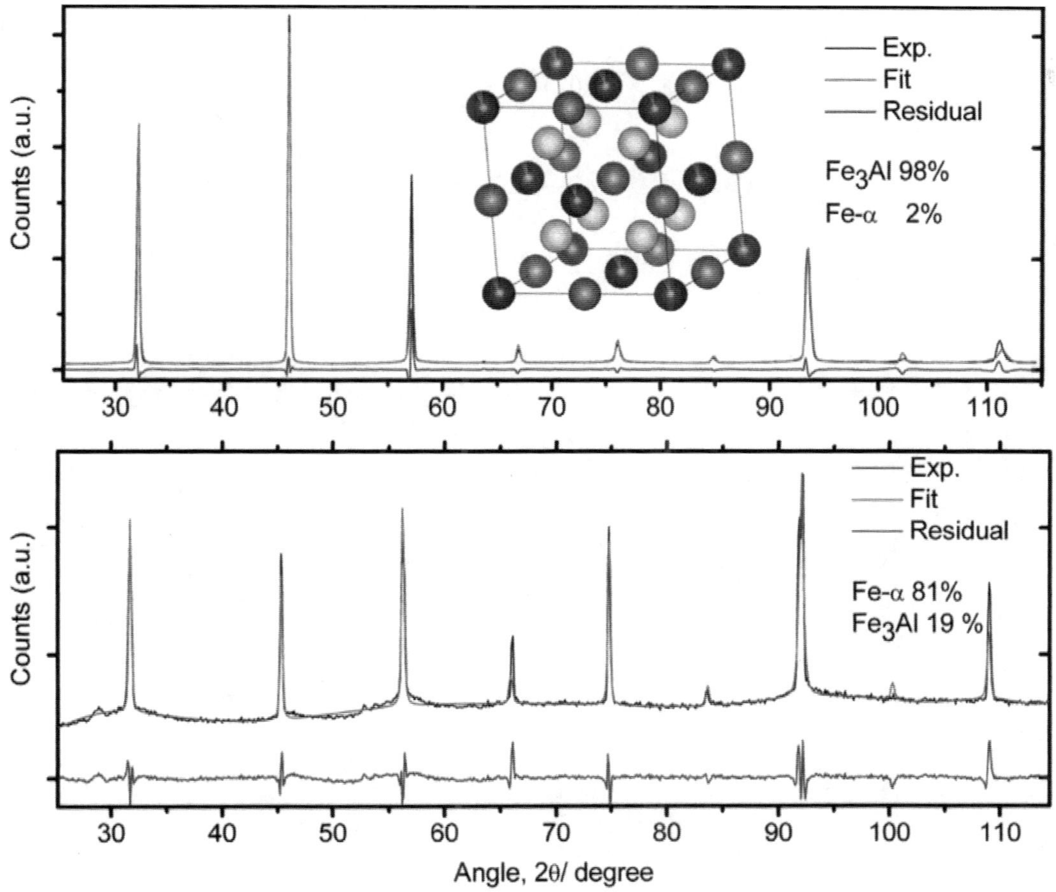

Figure 2. Rietveld refinement of the diffraction patterns at 300 K (top) and 1200 K (bottom). The top panel also shows the unit cell of the Fe_3Al-type structure, where blue and fuchsia spheres represent Fe atoms, red ones Ge atoms and turquoise ones Fe or Al atoms.

concentration in the sample is 8 at.%, it is impossible to detect an ordered structure from these reflections. Bearing this in mind, the presence of an ordered phase was detected by inspecting the peak splitting at high angles.

The Fig. 1 (right panel) shows the diffracted peaks at around 92° for temperatures between 300 K and 1200 K. The temperature associated to each diffractogram corresponds to the mean temperature of the interval in which the averages were performed. At high temperature we observe two reflections at around 92°, whereas at room temperature only one reflection (around 93.4°) appears. The presence of two peaks at high temperature is associated with the presence of two crystalline structures. As the temperature decreases, the peak associated with Fe-α decreases at the expenses of the other. In order to model this behavior, we tested different structures involving Fe, Al and Ge and describing the diffracted patterns. The best fitting model for data at room temperature is composed of a Fe_3Al-type structure with atoms at the positions and with the occupations shown in Table 1.

The Fig. 2 shows the final Rietveld refinements, where we observe that the strong texture effects could not be completely modeled with spherical harmonic functions (top panel), thus the fit goodness is quite poor (Rwp factor around 10%). Nevertheless, the lattice parameters of the Fe-α and Fe_3Al phases were accurately obtained, resulting in (2.923 ± 0.003) Å for Fe-α at

Element	Position	% Occ. factor
Fe	$4a$ $(0, 0, 0)$	86
Ge	$4a$ $(0, 0, 0)$	14
Fe	$4b$ $(1/2, 1/2, 1/2)$	100
Fe	$8c$ $(1/4, 1/4, 1/4)$	97
Al	$8c$ $(1/4, 1/4, 1/4)$	3

Table 1. Crystallographic data of the Fe_3Al-type structure ($Fm\bar{3}m$) used to model the diffractograms at 300 K.

1200 K and (5.752 ± 0.002) Å for Fe_3Al at 300 K. The bottom panel shows the high-temperature difractogram, where we observe a coexistence of the ordered (Fe_3Al type) and disordered (Fe-α type) phases, the latter being predominant [16].

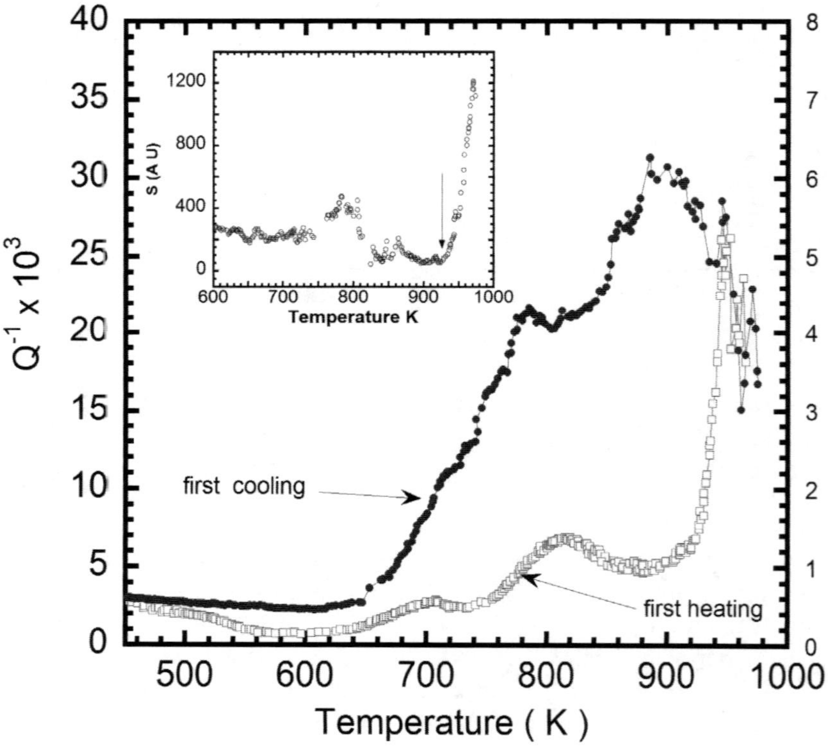

Figure 3. Mechanical spectroscopy, first thermal cycle to 973 K. Empty squares heating, full circles cooling. The inset shows the strength of the amplitude-dependent damping behavior on heating up to 973 K, and the arrow indicates the order decrease.

Damping measurements exhibit a hysteresis behavior between the heating run performed up to temperatures over 950 K and the subsequent cooling, see the Fig. 3, which is in agreement with the order evolution shown above from NPD studies [8, 9, 10, 11, 12]. In fact, the decrease in the order degree above 950 K leads to an increase in the dislocation mobility giving rise to the recovery of quenched-in dislocations at elevated temperatures. Thus, during the cooling run, the height of grain boundaries peaks related to the solvent and solute [14, 15] increases due to

Journal of Physics: Conference Series **663** (2015) 012004 doi:10.1088/1742-6596/663/1/012004

the reorganization of the structure [8, 9, 10, 11, 12]. In addition, the strength of amplitude-dependent damping (S) shows a change in slope at around the temperature where the order degree markedly decreases (see inset in Fig. 3) due to the increase in the dislocation mobility. The behavior of S as a function of temperature is also in agreement with the evolution of the order shown above and with previous works [8].

4. Conclusions

We determined the structure of Fe - 8at.%Al - 4at.%Ge using the neutron thermo-diffraction technique. At room temperature we found a Fe_3Al-type ordered structure with a deficiency of Al in the $8c$ sites. The Ge atoms are distributed in the $4a$ and Al atoms in $8c$ sites and we observe an order-disorder transformation at around 950 K when the crystal structure becomes an Fe-α type.

The disappearance of the order over 950 K gives rise to the hysteresis behavior of damping between the heating and the cooling runs, when the sample is heated over the phase transition temperature. The strength of the amplitude-dependent damping clearly reveals that at 950 K an order-decrease process is taking place.

Acknowledgements

The Institut Laue Langevin (ILL) is acknowledged for the allocated neutron beam time (Exp. 1-1-94), and D20 team for the help conducting the experiments. This work was partially supported by the CONICET, PIP 0179, and the PID-UNR; ING 290 and 453. The authors also want to thanks Prof Igor Golovin for supplying the samples.

References

[1] Bozorth R M, 1951 Ferromagnetism, MacMillan, London.
[2] Brandes E A, Brook G B, 1999 Smithells Metals Reference Book, Butterworth/Heinemann, Oxford.
[3] Cullity B D,1972 Introduction to Magnetic Materials, Adison Wesley, Reading.
[4] Raghavan V, 2002 J. Phase Eq. **23**, 437.
[5] Golovin I S, Strahl A, Neuhäuser H, 2006 Int. J. Mater. Res. **97**, 1078.
[6] Hansen T C, Henry P F, Fischer H E, Torregrossa J, Convert P, 2008 Meas. Sci. Technol. **19**, 034001.
[7] Rodríguez-Carvajal J, 1993, Physica B **192**, 55.
[8] Lambri O A, Pérez-Landazábal J I, Gargicevich D, Recarte V, Bonifacich F G, Cuello G J, Sánchez-Alarcos V, 2015, Mater. Trans. JIM **56**, 182.
[9] Lambri O A, Pérez-Landazábal J I, Cano J A, Recarte V, 2004, Mater. Sci. Eng. A **370**, 459.
[10] Lambri O A, Pérez-Landazábal J I, Cuello G J, Cano J A, Recarte V, Siemers C, Golovin I S, 2009, J. Alloys Compounds **468**, 96.
[11] Lambri O A, Pérez-Landazábal J I, Recarte V, Cuello G J, Cano J A, Golovin I S, 2012, J. Alloys Compounds **537**, 117.
[12] Lambri O A, Pérez-Landazábal J I, Cuello G J, Gargicevich D, Recarte V, Bonifacich F G, Giordano E D, Sánchez-Alarcos V, 2014, Neutron News **25**, 37.
[13] Lambri O A, 2000, in Materials Instabilities, ed. by J. Martínez-Mardones, D. Walgraef and C. H. Wörner, (World Scientific, New Jersey, 2000) pp. 249-280
[14] Schaller R, Fantozzi G, Gremaud G (Eds.), 2001 Mechanical Spectroscopy Q^{-1}, Trans. Technical Publication, Uetikon-Zuerich Switzerland.
[15] Nowick A S, Berry B S, 1972 Anelastic Relaxation in Crystalline Solids, Academic Press New York
[16] Momma K, Izumi F, 2011, J. Appl. Crystallogr. **44**, 1272.

Journal of Physics: Conference Series **663** (2015) 012005 doi:10.1088/1742-6596/663/1/012005

Structural effects of the spin-state crossover at high temperature in the distorted $ErCoO_3$ cobaltite

Jessica Padilla-Pantoja[1], José Luis García-Muñoz[1,*], J. A. Alonso[2], M T Fernandez-Díaz[3]

[1] Institut de Ciència de Materials de Barcelona, ICMAB-CSIC, Campus Universitari de Bellaterra, E-08193, Bellaterra, Barcelona, Spain.
[2] Instituto de ciencia de Materiales de Madrid, CSIC, Cantoblanco, E-28049, Madrid, Spain.
[3] Institut Laue Langevin, BP 156X, Grenoble, F-38042, France.

[*] Email: garcia.munoz@icmab.es

Abstract. We present a neutron powder diffraction study reporting structural evidences of a spin-state crossover beginning at T_{SS}≈575 K in the highly distorted $ErCoO_3$ cobaltite, based on measurements in the temperature range 100 K - 1000 K. Different interrelated changes in the structural evolution upon heating have been described, which are driven by an anomalous expansion of the octahedra. The activation of IS or HS spin states brings about an expansion of the CoO_6 octahedra (increase of Co-O bond lengths), and the augmentation of the Co-O distances in the *a-b* plane respect to that along the c-axis. The spin-state crossover extends over a broad temperature interval and can be monitored by an atypical augmentation of the orthorhombic strain.

1. Introduction

The spin-state nature across the two electronic transitions in $LaCoO_3$ has been matter of controversy for many years [1-8]. $LaCoO_3$ exhibits two broad transitions: there is first a spin-state crossover (SSC) at T_{SS} (between ~50-100 K), where a hump in the magnetic susceptibility signals the appearance of a paramagnetic semiconductor. The second one is an insulator-metal transition (T_{IM}) between ~500-600 K. The stabilization of one of the three spin-states in octahedral coordination depends on the ratio between the crystal field and Hund's rule coupling. For large splitting all the electrons occupy the low energy t_{2g}-levels resulting in zero spin and orbital momentum (low spin state, LS). States with finite spin momentum arise from the transfer of one or two electrons into the higher energy e_g-orbitals and correspond to an intermediate spin (IS) or high spin (HS) state, respectively. So, the two most often investigated pictures in $LaCoO_3$ involve the spin transition either LS→(HS+LS)→IS [1,2] or LS→IS→(HS+IS) [3-6]. However, in presence of disorder phenomena, a LS→HS crossover at T_{SS} can not be totally ruled out [2,7].

In addition, the ground and thermally excited states of $Ln$$CoO_3$ perovskites (*Ln*:lanthanide or rare earth) with heavy rare-earths and more distorted structures are also attracting great interest [5,8-11]. So, some reports evidenced that upon increasing temperature two separate electronic transitions occur as well in the perovskites of this series with smaller volume. Decreasing the size of *Ln* atoms the excited states would be occupied at temperatures progressively higher. Furthermore, although there are evidences of a high temperature metallic phase in the $Ln$$CoO_3$ series even with lanthanides smaller than

La, Pr or Nd [8], the nature of the high temperature state is not well understood. The main purpose of the present paper was to investigate the lattice and electronic thermal evolution of $ErCoO_3$, a very distorted specimen of this series, putting emphasis on the structural and lattice evolution as an effective way to approach the electronic phenomena associated with the spin-state of Co^{3+} ions in the system.

2. Experimental

Polycrystalline $ErCoO_3$ was prepared by sol-gel method [12]. Stoichiometric amounts of Er_2O_3 and $Co(NO_3)_2.6H_2O$ were dissolved in citric acid, by adding HNO_3 to favour the solution of rare-earth oxide. The solution was slowly evaporated leading to a resin, which was first dried at 393K and then slowly decomposed at temperatures up to 873 K for 12 h. Finally, the obtained precursor was heated at 1173 K under a 200 bar oxygen pressure for 12h. Then, the sample was cooled down at 5 K/min to room temperature in order to obtain a good oxygenation.

X-ray diffraction patterns at RT were recorded using a Brucker D8 diffractometer (40 kV, 30 mA) and Cu Kα radiation. We found the sample to be single phased and without impurities. Neutron powder-diffraction (NPD) measurements were carried out on the high-resolution D2B diffractometer of the Institute Laue-Langevin (ILL, Grenoble) at temperatures ranging from 100 K to 973 K. Above room temperature the sample was warmed in a furnace from 300 K to 973 K, while measurements at lower temperatures were performed using a CCR cryostat. Due to the relatively large absorption-cross section for neutrons of Er and Co, a narrow 5mm sample-holder was used. NPD data were recorded at the neutron wavelength λ=1.594 Å and were analyzed with the Rietveld method using the FULLPROF suite of programs [13].

3. Results and discussion

All the NPD diffraction patterns for $ErCoO_3$ collected between 100 K and 973 K were successfully refined using a *Pbnm* (no. 62) space group with cell parameters $a \approx \sqrt{2}a_0$, $b \approx \sqrt{2}a_0$, $c \approx 2\ a_0$ (where a_0 is the ideal cubic parameter). As an example, Figure 1 shows the refined pattern measured at 600 K. The details of the structural model obtained from the refinement at this temperature are reported in Table I, together with the agreement factors.

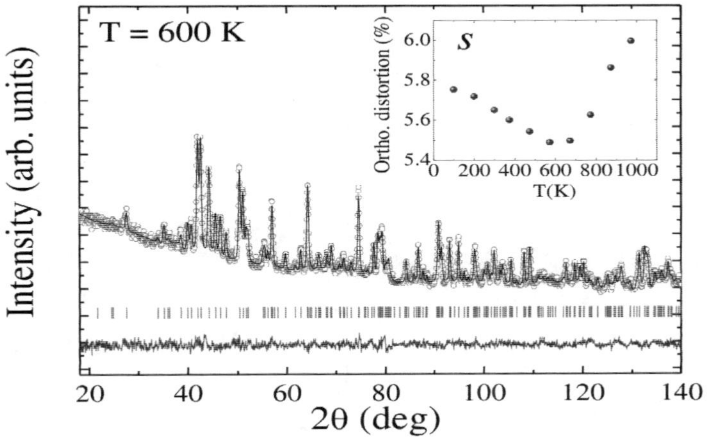

Figure 1. Rietveld refinement (black line) of neutron diffraction pattern from D2B (red circles, experimental points; bottom blue line, difference; green lines, Bragg reflections) for $ErCoO_3$ at 600 K. The inset shows the temperature dependence of the orthorhombic strain (*s*).

Table I. Structure at 600 K refined using SG *Pbnm*. Lattice parameters are a=5.1472(7) Å, b = 5.4379(9) Å, c = 7.3815(4) Å. Agreement parameters are χ^2 = 1.14 and R_B (%) = 6.36

Atom	Site	*x/a*	*y/b*	*z/c*	B_{iso} (Å^2)
Er	*4c*	0.9810(6)	0.0682(5)	0.2500	1.08(6)
Co	*4b*	0.5000	0.0000	0.0000	0.91(2)
O1	*4c*	0.0959(7)	0.4733(6)	0.2500	1.28(5)
O2	*8d*	0.6926(5)	0.3010(5)	0.0475(3)	1.30(5)

In Figure 2(a) we have plotted the thermal evolution of the cell parameters up to ~1000 K. Within the resolution of the technique, we did not detect the formation of oxygen vacancies up to the maximum temperature reached. In the figure it is shown that the relation between lattice parameters $b > c/\sqrt{2} > a$ (O type) expected for small rare-earths [8] remains in the whole temperature range. This comparison between parameters differs from the relation $a > c/\sqrt{2} > b$ characteristic of the $LnCoO_3$ cobaltites with larger rare-earths [9]. But the main feature to be underlined in Figure 2(a) is a remarkable extra expansion of the cell parameters starting around 575 K, which confirms a deviation on heating from the conventional thermal lattice expansion. The anomalous contribution to the lattice above that temperature is visible in the evolution of all three lattice constants represented in the Figure 2(a), but it is particularly apparent in the b(T) curve. Hence we present in Figure 2(b) the linear thermal expansion coefficient α_b deduced from b(T) as $\alpha_b = 1/b_0 \times \Delta b/\Delta T$. In Figure 2(b) the steep enhancement of the expansion coefficient α_b at T_{SS} is preceded by the regime where only phonons contribute to the thermal expansion. In this initial regime we observe a practically linear increase of the α_b expansion coefficient in the interval 100-550 K. Moreover, in the inset of the same figure we have plotted the temperature evolution of the unit cell volume. The solid red line, fitting the equation $V_0 = V_1 + V_2\,\Theta\coth(\Theta/T)$ before T_{SS}, is a guide to the eye.

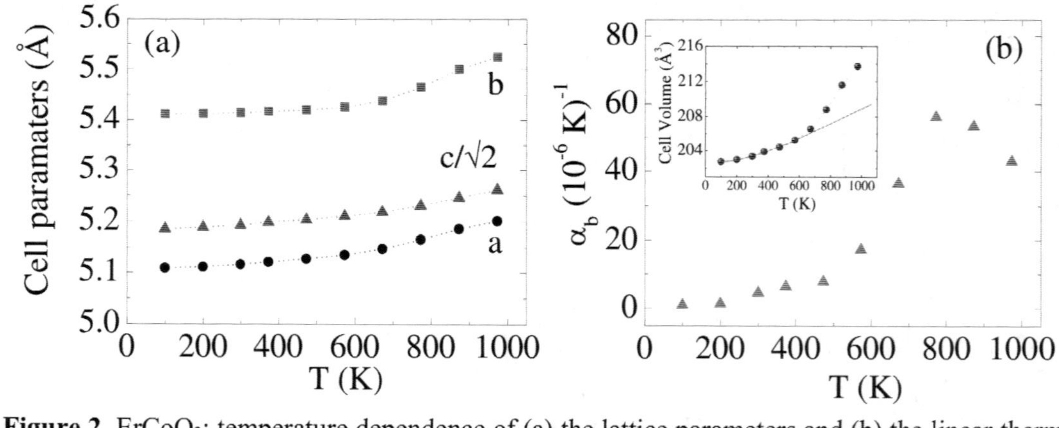

Figure 2. $ErCoO_3$: temperature dependence of (a) the lattice parameters and (b) the linear thermal expansion along b axis (α_b). The inset shows the temperature dependence of the cell volume for $ErCoO_3$. The solid red line, fitting the equation $V_0 = V_1 + V_2\,\Theta\coth(\Theta/T)$, is a guide to the eye.

The anomalous expansion shown in these figures above $T_{SS} \approx 575$ K is the signature of a sudden increase of populated excited magnetic states (IS or HS) in the Co^{3+} ions. The occupation of excited magnetic states is activated at $T_{SS} \approx 575$ K but it develops progressively along the interval 575-780 K. Above ~800 K the expansion coefficient α_b starts to decrease. From the inset of Figure 2(b) a rough estimation of the anomalous volume expansion accumulated at ~1000 K due to spin state excitations gives ~2 %.

Journal of Physics: Conference Series **663** (2015) 012005 doi:10.1088/1742-6596/663/1/012005

The spontaneous orthorhombic strain, defined as s=2(b-a)/(a+b), is plotted in the inset of Figure 1. It decreases on heating up to T_{SS}, where it presents a minimum. Above that temperature, the greater expansion of *b* (in comparison with $c/\sqrt{2}$ or *a*) preserves the relation $b > c/\sqrt{2} > a$ and increases *s*. Hence, in this system the occupation of higher spin states produces an enhancement of the orthorhombic distortion. In addition, we must observe in the same figure that *s* is still increasing at 973 K (our highest temperature reached), indicating that the spin state change process has not come to the end at that temperature.

The same behavior has been also observed in compounds with small rare earths such as $YCoO_3$, $DyCoO_3$ and $GdCoO_3$ [8]. The analysis of thermal expansion in the *b* lattice (α_b) as a function of temperature is also shown in Figure 2(b). We observe the expansion coefficient α presents an inflection point just at T_{SS} and it reaches a maximum at T~800 K where the spin-state transition could be completed. In contrast to larger *Ln* in which two shoulders are visible in the expansion coefficient, here only one maximum is detected.

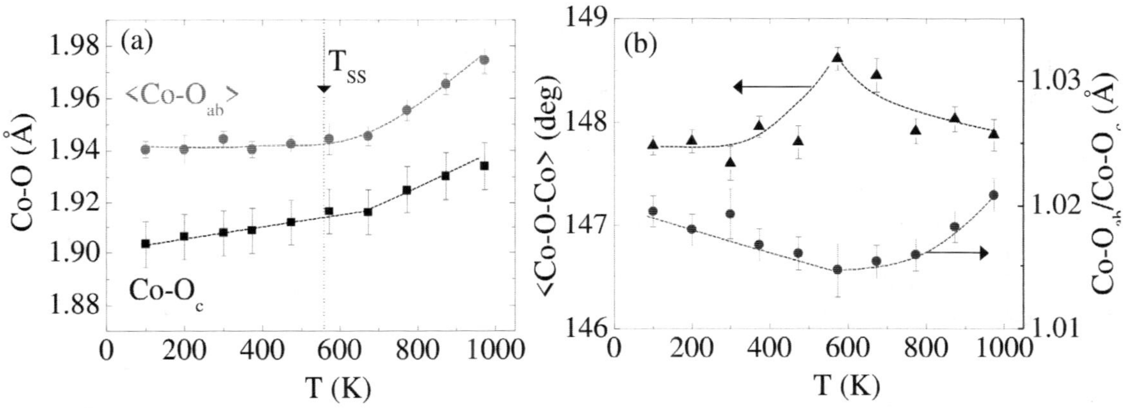

Figure 3. $ErCoO_3$: evolution with temperature of (a) the average Co-O_c (black squares) and Co-O_{ab} (red circles) bond lengths in the CoO_6 octahedra; (b) the average Co-O-Co bond angles (black triangles, left axis) and Co-O_{ab}/Co-O_c ratio (blue circles, right axis). The dotted lines are eye-guides.

The refined Co-O bond lengths in the equatorial plane (<Co-O_{ab}>) and apical axis of the CoO_6 octahedra are plotted in Figure 3(a). The average <Co-O-Co> bond angle for $ErCoO_3$ up to ~1000 K are displayed in Figure 3(b). The average of the Co-O distances in the *a-b* plane is larger than along the c-axis (Co-O_c) in the wide temperature range below 1000 K. However, the difference between them does not show a monotonous evolution: it is ~1.9% at 100 K, decreases under heating, reaches a minimum value of 1.5% around 575-600 K, and then increases again upon heating up to 2.1% at ~1000 K. Both distances exhibit a sudden enhancement at T_{SS} but it is steeper for the Co-O equatorial bonds. Therefore, the activation of IS or HS spin states brings about: (i) an expansion of the CoO_6 octahedra (increase of Co-O bond lengths) and (ii) the augmentation of the Co-O distances in the *a-b* plane respect to that along the c-axis. This resultant flattening of the octahedra due to spin state excitations is superimposed to the expected thermal diminution of the <Co-O_{ab}>/Co-O_c ratio due to the phonon (lattice) contribution.

On the other hand, the temperature evolution of the average <Co-O-Co> bond angle (see Figure 3(b), left axis) also reveals a non monotonous evolution. This angle displays a maximum around ~580 K coinciding with T_{SS}. The maximum value of the average bond angle (~148.5°) is very small due to the small size of the Er atoms. Contrarily to the conventional lattice evolution of distorted perovskites, we have observed a drop of the <Co-O-Co> bond angle in the range where the excitation to a higher spin state takes place. This unusual evolution must be considered as another expression of the enhancement of the orthorhombic distortion concomitant to the spin state change previously described. It is driven by the expansion in the CoO_6 octahedra observed upon heating above T_{SS}. To be noted is that the evolution becomes almost flat above 800 K. This evolution presents similarities with the evolution of the average bond angle previously reported in large $Ln$$CoO_3$ perovskites (Ln =La, Pr, Nd) above the second SSC [9]. The drop is much more pronounced for Er (small rare-earth) and hardly perceptible for La (the largest lanthanide).

4. Conclusion

The main purpose of this paper was to obtain experimental evidence of the electronic phenomena associated with the spin-state of Co^{3+} ions through the characterization of the structural and lattice evolution up to very high temperatures. To our knowledge, $ErCoO_3$ is the smallest Co^{3+} simple perovskite investigated so far in a broad temperature range up to ~1000K. We have observed one shoulder or clear maxima in the thermal evolution of the linear expansion coefficient resulting from the activation of a broad spin-state crossover from LS to IS or HS states that begins at T_{SS}≈575 K in this highly distorted cobaltite. Several interrelated structural changes above T_{SS} are driven by the anomalous expansion of the octahedra. The activation of IS or HS spin states brings about an expansion of the CoO_6 octahedra, and the augmentation of the Co-O distances in the *a-b* plane respect to that along the *c*-axis. The spin-state crossover extends over a broad temperature interval and can be monitored by an atypical augmentation of the orthorhombic strain. Further experimental work is needed to clarify the nature of the spin transitions in $Ln$$CoO_3$ cobaltites with, respectively, heavy and light lanthanides.

Acknowledgements
We thank financial support from MINECO (Spanish government) under projects MAT2009-09308, MAT2012-38213-C02-02 and MAT2013-41099-R, and to the ILL for granting beamtime. J. P.-P. thanks the CSIC for a JAE-Predoc contract.

References

[1] Yamaguchi S, Okimoto Y, Taniguchi H, and Tokura Y 1996 *Phys. Rev. B* **53** R2926
[2] Podlesnyak A, Streule S, Mesot J, Medarde M, Pomjakushina E, Conder K, Tanaka A, Haverkort M W, and Khomskii D I 2006 *Phys. Rev. Lett.* **97** 247208
[3] Asai K, Yoneda A, Yokokura O, Tranquada J M and Shirane G J 1998 *Phys. Soc. Jpn.* **67** 290
[4] Klie R F, Zheng J C, Zhu Y, Varela M, Wu J and Leighton C 2007 *Phys. Rev. Lett.* **99** 047203
[5] Nekrasov I A, Streltsov S V, Korotin M A, and Anisimov V I 2003 *Phys. Rev. B* **68** 235113
[6] Louca D and Sarrao J L 2003 *Phys. Rev. Lett.* **91** 155501
[7] Haverkort M W, Hu Z, Cezar J C, Burnus T, Hartmann H, Reuther M, Zobel C, Lorenz T, Tanaka A, Brookes N B, Hsieh H H, Lin H-J, Chen C T and Tjeng L H 2006 *Phys. Rev. Lett.* **97**176405
[8] Knížek K, Jirák Z, Hejtmánek J, Veverka M, Maryško M, Maris G and Palstra T T M 2005 *Eur. Phys. J. B* **47** 213
[9] Knížek K, Hejtmánek J, Jirák Z, Tomeš P, Henry P and André G 2009 *Phys. Rev. B* **79** 134103
[10] Yu J, Phelan D, and Louca D 2011 *Phys. Rev. B* **84** 132410
[11] Zhang G, Gorelov E, Koch E and Pavarini E 2012 *Phys. Rev. B* **86** 184413
[12] Alonso J A, Martínez-Lope M J, de la Calle C, and Pomjakushin V 2006 *J. Mater. Chem.* **16** 1555
[13] Rodríguez-Carvajal J 1993 *Physica B* **192** 55

Journal of Physics: Conference Series **663** (2015) 012006 doi:10.1088/1742-6596/663/1/012006

Crystallization of *para*-Hydrogen: a quantum phase transition at finite temperature?

C Cabrillo[1], F Fernández-Alonso[2], R Fernández-Perea[1], F J Bermejo[1], M A González[3], C Mondelli[3] and E Farhi[3]

[1] Instituto de Estructura de la Materia, IEM-CSIC, Serrano 123, Madrid 28006, Spain.
[2] ISIS Facility, Rutherford Appleton Laboratory, Chilton, Oxfordshire OX11 0QX, U. K.
[3] Institut Laue Langevin, 6 rue Jules Horowitz 38042, Grenoble Cedex 9, France

E-mail: ccabrilo@foton0.iem.csic.es

Abstract. Recent neutron scattering measurements have revealed that *para*-Hydrogen prepared in its fundamental state solidifies in a crystal that do not show temperature response (temperature independent molecular displacement and first rotational level linewidth as well as negligible thermal expansion). Here, it is argued that such unique behaviour in the crystal phase can be understood if the crystallization is driven by pure quantum fluctuations.

1. Introduction

Local density fluctuations triggers crystal growing within a liquid at the freezing temperature. Conversely, as Lindemann observed long time ago [1], the crystal melts when the root mean square (rms) displacement is a sizeable portion of the first neighbour distance. At that point, some particle displacements explore too much the core part of the interaction potential acquiring enough kinematic energy to scape the crystal ordering. That is, phase transitions are driven by fluctuations. Once the crystal emerges, fluctuations decrease as temperature decreases. Eventually, the rms displacement is small enough for the harmonic approximation to be valid. Unless that quantum delocalization imposes a minimum size of the fluctuations too large. That being so, even at the lowest achievable temperature the harmonic approximation breakdowns and the solid is considered a quantum solid. Such is the case of the hydrogen family of molecular solids [2]. As the temperature decreases eventually quantum fluctuations overcome the diminishing thermal contribution so that both the rms displacement and the thermal expansion eventually stop diminishing to accommodate to the finite value of quantum fluctuations. Recently we have discovered that this athermal behaviour is stretched to its maximum in the case of *para*-Hydrogen in its fundamental state ($J = 0$). Remarkably, the *para*-$H_2^{J=0}$ molecular crystal shows no discernible thermal expansion nor increasing rms displacement from $\sim$ 3 K to melting ($\sim$ 13.8 K) as attested by neutron scattering measurements of the $J = 0 \rightarrow 1$ transition linewidth, rms molecular displacement and Bragg scattering [3]. As we all know, in Helium the effect is even stronger. Quantum fluctuations are so large that at its saturated vapour pressure Lindemann melting criterion is fulfilled even at zero temperature: at the lowest temperatures the condensate phase remains liquid (a quantum liquid). The survival of fluctuations, albeit quantum, immediately opens the possibility of phase transitions even at $T = 0$ K. Necessarily driven by quantum fluctuations they are known as Quantum Phase

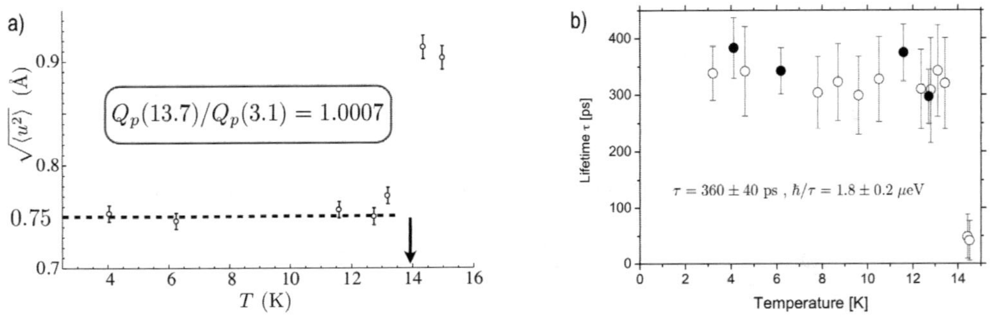

Figure 1. a) White circles are the rms of molecular displacements of the *para*-$H_2^{J=0}$ crystal at various temperatures. The arrow marks the triple point temperature (13.8 K). The blue box display the ratio of Q corresponding to the $(\bar{1}02)$ reflection of the *hcp* crystal structure (denoted by Q_p), at 13.7 K and 3.1 K. b) The temperature dependence of the lifetime of the $J = 0 \rightarrow 1$ transition. The average value and its corresponding value in μeV are also given.

Transitions (QPTs), a field of considerable interest in the last two decades [4]. Indeed, ^{4}He below 1 K undergoes a first order transition towards a hcp crystal at around a pressure of 25 bar. Second order QPTs are also observed and, as in the classical case, are far more fashionable. Here, we will argue that the very peculiar athermal behaviour of the *para*-$H_2^{J=0}$ crystal can be understood if the corresponding crystallization is driven by fluctuations which are quantum in nature in spite of being at a finite temperature. As such, it can be considered a QPT between a so called quantum Boltzmann liquid and a "high temperature" quantum crystal.

2. Experimental results

The main goal of the experiment above mentioned was to measure the spectral broadening of the first $J = 0 \rightarrow 1$ rotational transition in solid *para*-H_2. Previous to the neutron scattering measurements a $J = 0$ sample was prepared carefully by passing the liquid H_2 through cromium oxide within a gel during five days. All measurements were performed on the IN20 thermal neutron spectrometer at the Institut Laue Langevin, France. The instrument was used either in polarised triple-axis (TAS) mode or in TAS-spin-echo (TASSE) mode, in both cases with fixed final wave vector (2.66 Å^{-1}). In TASSE mode, the accessible Fourier time spanned 3.2 ns at ~2.7 Å^{-1} with a time constant of 440 ps yielding an energy resolution of 1.15 μeV half width at half maximum at ~15 meV. We not only measured the sought for linewidth but Bragg diffraction and the rms displacement in a temperature range between ~ 3 K and the triple point at 13.8 K. The results are summarized on Fig. 1. For the details see [3]. Here, our interest is in the fact that the crystal show an essentially athermal behaviour, as it is illustrated in the Figure.

3. Discussion

Probably the only indisputable fact coming out from the result referred to above is that the *para*-H_2 cryocrystal cannot be described classically. A full quantum calculation of the system is unaffordable. We can, however, eliminate the rotational degrees of freedom since we are dealing with a $J = 0$ population of diatomic molecules with a first rotational level at ~ 171 K while the system temperature is below 20 K. Silvera and Goldman [5] developed an accurate effective pair potential for the $H_2^{J=0}$ molecule some time ago. Using this potential (isotropic as it corresponds to a $J = 0$ wavefunction) we have performed a classical Molecular Dynamics study

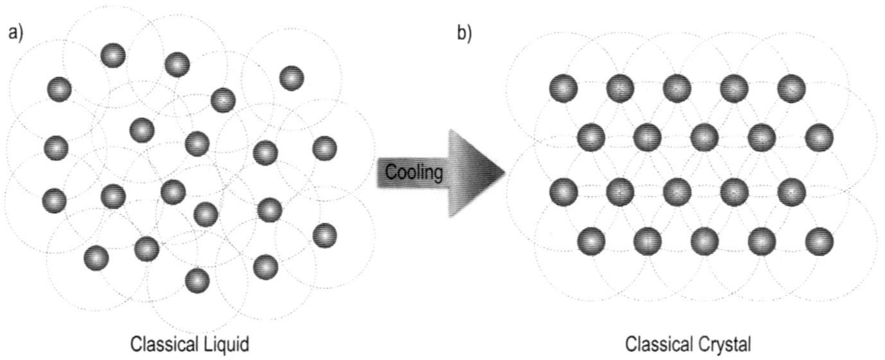

Figure 2. Crystallization of a classical 2D crystal. In a) temperature is near but above T_{tp}. In b) the crystal is shown at $T = 0$.

of the condensate phases. This classical analog melts at $T_{tpcl} \sim 25$ K, around 1.8 times higher than the experimental value (at any rate, consistent with classical law-of-corresponding-states arguments). Scaling to the respective triple point temperature the corresponding calculated $S(Q)$'s yield a ratio $Q_p(0.95\ T_{tpcl})/Q_p(0.22\ T_{tpcl})$ that amounts to 1.04 to be compared with the measured value of 1.0007. Finally, the classical crystal show a temperature dependent rms displacement. However, Lindemann's argument is vindicated (and subsidiarily Silvera-Goldman potential) since it melts at a $\sqrt{\langle u^2 \rangle} \sim 0.75$ Å in agreement with the measurements.

As expected, the classical analog cannot reproduces the observed behaviour but, since it melts at the correct $\sqrt{\langle u^2 \rangle}$, we gain a strong confidence in the adequacy of the Silvera-Goldman potential. Under such an assumption, the classical analog tell us that quantum delocalization starts to be relevant somewhere above 25 K, i.e., very well above the triple point. This is consistent with a good deal of previous observations regarding liquid *para*-H_2. The excess of kinetic energy arising from the Heisenberg uncertainty Principle has been observed [6, 7] and it results in a remarkable isothermal dependence of the kinetic energy on density [8]. As in the case of $\sqrt{\langle u^2 \rangle}$ in excess in a quantum crystal, this phenomenology can be understood in terms of single-molecule wavefunction exploring the repulsive core of the intermolecular potential. Such a wave mechanics behaviour explain also why the liquid avoid crystallization at 25 K. Similarly, owing to the effects of particle delocalization on the overall magnitude of attractive intermolecular interactions, liquid *para*-H_2 near its triple point can sustain collective excitations which are otherwise absent in the classical computational (supercooled) counterpart [9]. All things considered, liquid *para*-H_2 qualifies as a quantum liquid. Since it crystallizes before Bose-Einstein condensation takes place, the term Boltzmann quantum liquid has been coined.

Being the quantum effects so manifest from so high temperatures, we are induced to consider what are the consequences of a fully developed wavelike behaviour already in the liquid, i.e., the consequences of classical thermal fluctuations playing no role. At a first glance, crystallization would be avoided altogether as in the case of He. However, excluded volume arguments make clear that it is not necessarily the case. To illustrate the point we will recourse to a 2D pictorial representation. Fig. 2 represents a classical picture of a 2D simple liquid crystallization. There, the blue spheres represents the core of the potential while the dotted line circle corresponds to the minimum of the pair potential. In the liquid side, the kinetic energy is too high for the crystal to form. Below T_{tp}, it is more energetically favorable for the particles to remain around the attractive well of the pair potential. At the transition, the excess of kinetic energy is released as heat more than compensating the diminution of configurational entropy. At sufficiently low temperature, the particles moves around the minima in the potential energy and the harmonic

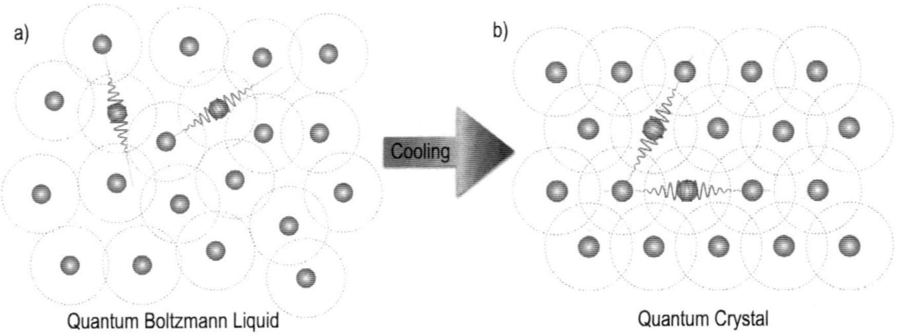

Figure 3. Crystallization of a quantum Boltzmann 2D liquid. In a) temperature is near but above T_{tp}. In b) the representation is valid for any temperature below T_{tp}.

approximation works. At $T = 0$ the particles sit at the minima as shown in Fig. 2 b).

Figure 3 represents the case of a quantum Boltzmann liquid fully dominated by quantum fluctuations (blue wavelets represent wavepackets). The scale of the quantum delocalization is commensurable with the interparticle distance. In fact, the wavepacket width determines the equilibrium density in order to avoid a substantial penetration of the wavefunctions within the repulsive core of the neighbour particles. Excluded volume around each particle is then maximized at the scale of the wavepacket width, not to the size of the repulsive core. If the well of the pair potential is shallow enough, decreasing the temperature would never drives the system to a crystalline phase. The balance between kinetic and potential energy will remains always in the kinetic side. But the configurations maximizing the available volume per particle are ordered close-packed structures (the triangular lattice in the 2D case). Furthermore, there is a gap in the available volume per particle between a close-packed arrangement and any other disordered configuration. Therefore, for a given density, by adopting a close-packed configuration, the system minimizes the wavepacket exploration of the neighbor particles core optimally: maximum first-neigbourg distance for each and every particle in the system. If it happens that for a given density the potential well is able to move the energy balance to the potential side for the close-packed configuration, the liquid will crystalize. But, under the assumption referred to above of no role of classical thermal fluctuations at the transition, decreasing the temperature do not change the picture and the crystal will remain fully expanded whatever the temperature. We conclude then, that the unique athermal behaviour observed in the *para*-$H_2^{J=0}$ crystal is a strong evidence in favour of a quantum nature of its crystallization transition.

Acknowledgments

Work supported in part by grants MAT2007-65711-C04-01 and MAT2012-33633 from the Spanish Ministerio de Economía y Competitividad.

References

[1] F A Lindemann 1910 Phys. Z. **11** 609
[2] I F Silvera 1980 Rev. Mod. Phys. **52** 393
[3] F Fernandez-Alonso et al 2012 Phys. Rev. B **86** 144524
[4] M Vojta 2003 Rep. Prog. Phys. **66** 2069
[5] I F Silvera and V V Goldman 1978 J. Chem. Phys. **69** 4209
[6] W Langel, D L Price, R O Simmons, and P E Sokol 1988 Phys. Rev. B **38** 11275
[7] M Celli, D Colognesi, and M Zoppi 2000 Eur. Phys. J. B **14** 239
[8] M Zoppi, D Colognesi, and M Celli 2001 Eur. Phys. J. B **23** 171
[9] F J Bermejo et al 2000 Phys. Rev. Lett. **84** 5359

Comments on Proton Cascade in a Molecular Solid: H/D Exchange on Mobile and Immobile Water

E. Forcén-Vázquez,[1,*,‡], L. R. Falvello, [1] M. Tomás[2]

[1]Dpto. de Química Inorgánica e I.C.M.A., U. de Zaragoza - C.S.I.C., Pedro Cerbuna 12, E-50009 (Spain) [2]Dpto. de Química Inorgánica e I.S.Q.C.H., U. de Zaragoza - C.S.I.C., Pedro Cerbuna 12, E-50009 (Spain) [‡]Current address: Dpto. de Física del Estado Sólido a Bajas Temperaturas. ICMA (C.S.I.C.) Pedro Cerbuna 12, E-50009 (Spain).

E-mail: elenafv@unizar.es

Abstract. Following a previous report of proton transport in a channel-free molecular solid possessing a complex set of water wires, the mechanisms implied by the available neutron diffraction measurements are further developed. Three specific proton transfer pathways along an experimentally observed water wire are proposed. The first is proton transfer involving two aqua ligands coordinated to the same Mn(II) unit and one free water acting as a shuttle. The second mechanism consists of proton transfer between coordinated aqua ligands of distinct molecules, with no shuttle. The third mechanism involves a component of vehicle transport, involving a water molecule that changes position during the process. Geometric parameters of the proposed mechanisms are given.

1. Introduction

Many biological and technological processes (*e.g.,* Adenosine triphosphate (ATP) synthesis in the former case or hydrogen fuel cells for the latter), require the presence of protons. Acid-base, redox and enzymatic reactions are based on proton transfer from one reagent to another. Proton mobility in bulk water is much higher, about 4.5 times greater than the mobility of other cations in different solvents, including water. Greater insight into the mechanism through which protons are transferred in bulk water can facilitate the understanding and development of other processes involving ionic movement.[1]

It has been observed in biological systems that the presence of a one dimensional H-bonded water chain (or "water wire" or "proton wire") facilitates proton transport through membranes. The associated transport mechanism, commonly called the Grotthuss mechanism,[2] is based on concepts put forward by Grotthuss in 1806 in the context of electrolysis. The mechanism, as conceived for proton transport, consists of the movement of an excess proton along the H-bonded water chain while the oxygen atoms of the water molecules remain in place. Materials showing this behaviour are known as proton conductors. Specific proposals involving Eigen and Zundel cations have been put forward for this process,[3] although there is still debate on the details.

A two-stage process is generally accepted as the basis of the Grotthuss mechanism of proton transport in liquid electrolytes and solids.[4] The process begins with the transfer of a proton from one molecule of water to an adjacent one, with both molecules being links in a water wire. This proton jump involves the rupture and formation of several covalent and hydrogen bonds in the neighbourhood of the molecules involved. Figure S1 represents the approach of a proton to the terminal water molecule of a chain (a).

The water molecule that binds the newly incorporated proton shares one of its H atoms with an adjacent water molecule, forming a Zundel dimer (b). As this shared proton becomes fully bonded to the O atom of the neighbouring water molecule, the latter in turn loses one of its original H atoms to the next water molecule along the chain (c). This process is repeated along the water chain and finally an H^+ exits the chain (b-d). This process inverts the dipole moments of the water molecules, and the original state is restored by reorientation of the molecules (e). Once the conformation of the water molecules is identical to that of the original state, the system can host a new proton jump (f).

Another feasible mechanism, the "vehicle mechanism," entails protons traveling on a host molecule. A host molecule, the carrier, is a species that can interact with protons, such as H_2O or NH_3, so the transfer takes places as an H_3O^+ or an NH_4^+, instead of as an H^+. The substance behaves as a Brønsted base and its diffusion coefficients are the limiting factors in proton conduction. The main difference here is that an H-bonded chain is not required and stronger basic species such as NH_3 can be used, which has helped for example in the preparation of new materials based on zeolites showing proton conductivity.[5]

2. Experimental

Full details of the measurements on which this discussion is based are given in Reference 6a and its accompanying Supporting Information.

3. Discussion

Compound **1** has been described in detail.[6] It is a 1-D coordination polymer that crystallizes in the monoclinic system, space group *P2₁/n*. The structural building block of the polymer is an $[Mn_4(citrate)_4]^{8-}$ cubane, which together with bridging $[Mn(H_2O)_4]_2^+$ units, forms a chain that is propagated along a crystallographic *n*-glide plane. Balancing the negative charge of the polymer are one pendant $[Mn(H_2O)_5]^{2+}$ unit coordinated to the periphery of each cubane and two freestanding $[Mn(H_2O)_6]^{2+}$ units between the polymer chains. In addition, there are nine free water sites per asymmetric unit, two of which are half-occupied. There are thus a total of 21 coordinated water molecules and, stoichiometrically, eight crystallization water molecules per unit of the polymer. The crystal is not porous and presents no channels in its structure. Nevertheless, when it is exposed to the atmosphere it undergoes a reversible dehydration in which three molecules of water (occupying four crystallographic sites) are lost, but with preservation of the crystallinity of the sample (compound **2**). There is a reduction of about 250 Å^3 in the volume of the unit cell, but the crystal still belongs to the monoclinic space group $P2_1/n$. When a crystal of **2** is left inside a closed vial, with a humid atmosphere created using moist cotton (among other methods), the initial derivative is fully restored (compound **1'**). This reversible process was characterized using single crystal X-ray and neutron diffraction (Figure S2).

As described previously,[6] when the rehydration is carried out using D_2O instead of H_2O, the structure of **1'** shows deuteration not only at the four crystallographic sites from which water egress takes place, but also at all of the remaining 26 water sites, including the 21 that are coordinated to Mn centers. This fact was definitively established using neutron diffraction, whose H/D isotope contrast provides irrefutable experimental characterization of the affected sites.

It is the purpose of the following discussion to expand on the mechanism that was proposed in Reference 6a, with more detail of the proton transport pathways that can operate in this crystal.

In compound **1**, as characterized by neutron diffraction, a significant H-atom disorder is observed for a number of the water molecules, with the 58 H atoms of the asymmetric unit occupying 72 crystallographic sites. The key to a cogent description of the mechanism through which such massive H/D substitution can take place in a non-porous molecular crystal, without loss of crystallinity, is the presence of an extensive H-bonded water structure: Several continuous motifs traverse a complete unit cell, with O...O distances of between 2.5 and 3.3Å. (Figure S3). As mentioned above, such a proton wire, or water wire, is a characteristic feature of proton conductors, and it is along such pathways that proton transport can take place.

Our original report[6a] described an H-bonded chain of water molecules presenting two different types of disorder (Figure S4). The first disorder pattern is present in eight of the 30 water fragments and consists of an oxygen atom [Ox (s.o.f. 1; s.o.f. = total site occupancy factor)] surrounded by three H atoms, one fully occupied [Hxa (s.o.f. 1)] and the remaining two with half occupancy each [Hxb and Hxc (s.o.f. 0.5)]. This can be split into two congeners with good geometry for a water fragment, Hxa-Ox-Hxb and Hxa-Ox-Hxc. The second type of disorder, found for two of the 30 sites, presents an oxygen atom attached to four half-occupied H sites [Hxa, Hxb, Hxc and Hxd (s.o.f. 0.5)]. This pattern can be split into two congeners, each with appropriate H-O distances and H-O-H angle for a water fragment, Hxa-Ox-Hxc and Hxb-Ox-Hxd. Each of the two congeners would be present half of the time. Proton movement along this 1-D chain is explained in terms of the Grotthuss mechanism. The proposed mechanism accounts for the splitting of each water site into two congeners as described above. One of these congeners, a, would be present in the first step, when the excess proton enters the chain (Figure S1). Note that the sites occupied by the hydrogen atoms in the b congeners are vacancies when congener a is present. The excess proton jumps from one water molecule to a second one, occupying a vacancy. This transfer is coordinated with the exit of one proton from the second water molecule; this proton is transferred to the next water molecule in the chain. Once protons have traversed the H-bonded water chain, the extra hydrogen leaves the chain. At this point, the dipole moment of the water molecules is inverted with respect to the starting state of the chain (b congeners). In order to restore the original situation and enable a new proton jump, a simple rotation and reorientation of the water molecules is needed.

Based on neutron diffraction results of the initial derivative, it was possible to identify three pathways through which protons were transferred from one molecule of water to an adjacent one. For these three situations, the experimental disorder observed for the H atoms matches the proposed mechanism, and a full description of the proton interchange can be given on the basis of the split of the disordered molecules into two congeners. All of the H-atom sites described in the following mechanisms were observed in the neutron diffraction analysis of compound **1**.

Mechanism 1: Proton transfer involving three water molecules, two coordinated to the same Mn(II) unit and one free water acting as a shuttle (Figure 1). This process involves O14w, attached to Mn(II) and surrounded by four half-occupied H sites; O30w, a crystallization water molecule; and the coordinated aqua O10w^{i}, both of the latter two having three H-atom sites, one fully occupied and the remaining two with half occupancies. In the initial arrangement, the congeners present are those formed by H14b-O14w-H14d, H30a(sof 1)-O30w-H30c(sof 0.5) and H10a^{i}(sof 1)-O10w^{i}-H10c^{i}(sof 0.5). The process starts with transfer of the half-occupied H14d to the H30b site bonded to O30w, a distance of 1.39(4) Å, Table S1. The distance H14d...O30w, 1.93(3) Å, is within the range considered acceptable for the Grotthuss mechanism. Concomitant with this proton jump, H30c hops to the site of H10b^{i} [distance: 0.73(3) Å] attached to O10w^{i} [H30c...O10w^{i} 1.74(4) Å]. At the same time H10c^{i} is transferred to another water molecule as will be described presently.

The loss of H14d at O14w is compensated by an incoming H atom that occupies the vacant H14c site. In the Grotthuss mechanism, proper geometry is maintained about O14w through rotation of this aqua ligand about the vicinal Mn7-O14w bond, which moves H14b to the H14a site.

To restore the initial situation and enable a new proton jump, some readjustments are needed. A rotation around the Mn7-O14w bond of about 90° would place H14a at the site of H14d, and of 104° will transform H14c into H14b. In a similar way, rotation of 109° around the H30a-O30w axis restores the initial H30a-O30w-H30c congener. Finally, a rotation of 110° around the H10a^{i}-O10w^{i} axis would move H10b^{i} to the site of H10c^{i}. This concerted set of rotations is the second step of the Grotthuss mechanism.

 doi:10.1088/1742-6596/663/1/012007

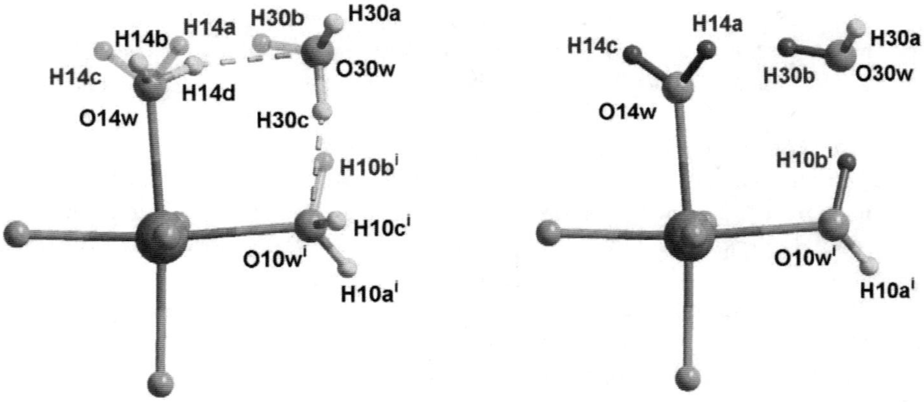

a)

Figure 1. First pathway: proton transfer between two water fragments coordinated to the same Mn(II) unit mediated by a free water molecule. a) Complete disorder observed around the water molecules. Atoms represented as solid spheres are present in the initial configuration. b) Final situation after the proton transfer.

Mechanism 2: Proton transfer between two water fragments coordinated to different Mn(II) units, with no intermediary. Figure 2 shows the disorder at the water sites involved in the transfer. This process, beginning with the H10a^{i}-O10w^{i}-H10c^{i} and H10a^{ii}-O10w^{ii}-H10b^{ii} congeners present, involves transfer of H10c^{i} to O10w^{ii} concomitant with the arrival of H10b^{i} as described for Mechanism 1 above. The aqua-complexes between which the proton jump occurs are related by a crystallographic inversion center. The H10c^{i} hopping distance to H10c^{ii} is 0.94(4) Å. As a consequence, H10b^{ii} exits O10w^{ii}, whose proton component now consists of H10a^{ii} and H10c^{ii}. After the proton jump, rotations around the H10a-O10w axes of the water molecules involved would restore the initial situation, moving H10b^{i} to the H10c^{i} site and H10c^{ii} at H10b^{ii}. A more complex development of this mechanism is possible, accounting for the bi-directionality implied by a crystallographic inversion center.

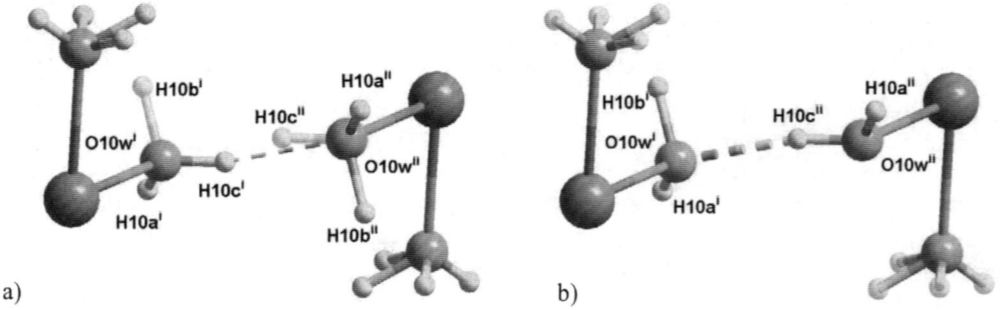

Figure 2. Second pathway: direct transfer between two water fragments coordinated to different Mn(II) units. a) Initial situation, only the congeners present before the proton jump are represented for O10w. b) Final situation after the proton jump between O10w^{i} and O10w^{ii}. i: -x, 1-y, -z; ii: 1+x, y, z.

Mechanism 3: Proton transfer between different Mn(II) units through an external water molecule acting as a shuttle. In this process the free water molecule, O28w, plays a crucial role. It is one of the water molecules that exit the sample when it is exposed to the air. Figure 3 shows the disorder in this region of the structure. O28w^{ii} is half occupied and is related by a crystallographic center of inversion at (1,1/2,0) with its congener O28w^{iii} at a distance of 1.94(3) Å. Several options can be considered to explain the observed disorder at the site of O28w. It could be that occupation of the two sites is randomly distributed throughout the solid; or it may be that the separate dispositions are required for proton movement in opposite directions in different regions of the crystal. Another possibility is that O28w changes its disposition in concert with the proton movement. This would impute the second mechanism for proton transport described in the Introduction, known as the vehicle mechanism.[7]

The proton movement observed in this third pathway is necessarily coordinated with the proton movements described in the first two mechanisms. First, at O14w^{ii}, H14c^{ii} jumps to O28w^{iii} [1.94(3) Å], coincident with the arrival of H30b^{iii} at the H14d^{ii} site on O14w^{ii} (Mechanism 1), as H14a^{ii} rotates to the H14b^{ii} site (H14b^{ii} and H14d^{ii} are semi-transparent in Figure 3a). Then, H28a^{iii} jumps to O14w^{iii}, occupying the site of H14c^{iii}. This second transfer is coordinated with the jump of H14d^{iii} to O30w, and the subsequent rotation of H14b^{iii} to the site of H14a^{iii}, with the final configuration at O14w^{iii} consisting of H14a^{iii} and H14c^{iii} atoms.

The rotations of H14b^{ii} and H14d^{ii} (Figure 3b) to the H14a^{ii} and H14c^{ii} positions (Figure 3a) and of H14a^{iii} and H14c^{iii} to the H14b^{iii} and H14d^{iii} sites leave O14w^{iii} ready to host a new hydrogen atom at the H14c^{iii} site and H14c^{ii} ready to jump in a subsequent cycle.

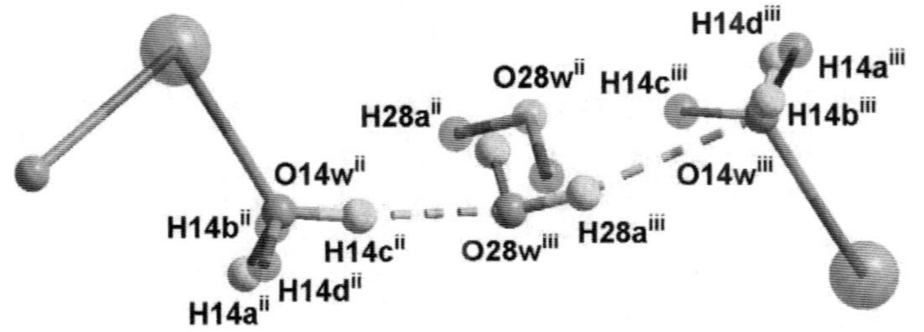

a)

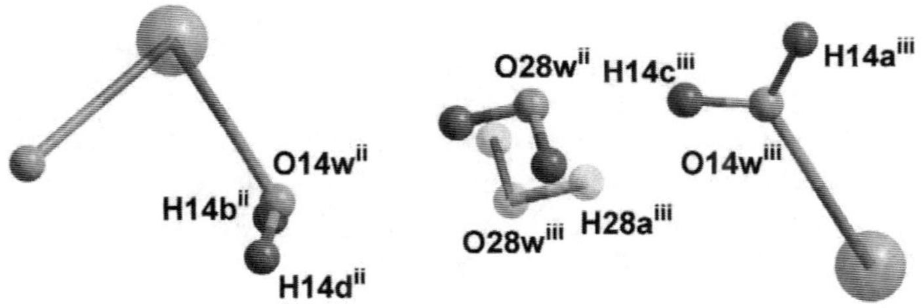

b)

Figure 3. Third pathway: Proton jump between two water molecules coordinated to different Mn(II) units. a) Experimentally observed disorder of the water molecules is shown. The atoms represented as solid spheres are present in the initial arrangement. b) Final disposition of the water molecules after the proton transfer. The two congeners of O28w are represented. i: -x, 1-y, -z; ii: 1+x, y, z; iii: 1-x, 1-y, -z .

The H...H jump distances for these proposed mechanisms are collected in Table S1, together with the distances from the initial H positions to the respective O atoms to which the protons are transferred. The jump distances range from 0.73(3) to 1.55(4) Å, with the larger distances corresponding to the pathway that involves O-atom motion.

4. Conclusions

Neutron diffraction reveals several pathways through which protons can be transferred between water fragments in compound **1**. The complexity of the mechanisms presents a challenge for constructing a fully detailed explanation of the proton movements and their coordination. What can be stated with certainty is that all water-bound H atoms, including those present on immobile waters bonded to Mn atoms, are substituted by deuterium in this non-porous molecular crystal. The presence of a complex system of H-bonded water wires provides an entry point to a description of possible proton-transfer mechanisms. The further presence of H-atom disorder, revealed accurately by neutron diffraction and compatible with the expected proton positions in a putative Grotthuss mechanism, permits a more detailed mechanistic description compatible with the experimental results and with a proposed proton transport protocol.

We propose mobile protons as the fundamental feature of this system, so it is important to emphasize that in experimental terms, with the diffraction measurements conducted at a temperature of 20 K, the protons are assumed to have occupied minimum-energy positions that permit this snapshot of what we conclude are stopping points in the Grotthuss mechanism.

References

[1] a) Kirchner B 2007 *Chemphyschem* **8** 41. b) Dellago C et al. 2003 *Physical Review Letters* **90**. c) Halding J. and Lomdahl P S 1988 *Physical Review A* **37** 2608.
[2] a) Nagle J F and Morowitz H J 1978 *Proc. Natl. Acad. Sci. USA* **75** 298. b) Grotthuss C J T 1806 *Ann. Chim.* **58** 54.
[3] Agmon N 1995 *Chemical Physics Letters* **244** 456.
[4] a) Markovitch H et al. 2008 *Journal of Physical Chemistry B* **112** 9456. b) Kornyshev A A et al. 2003 *Journal of Physical Chemistry B* **107** 3351. c) Day T J F et al. 2000 *Journal of the American Chemical Society* **122** 12027.
[5] Kreuer K D et al. 1982 *Angew. Chem. Int. Ed.* **21** 208.
[6] a) Capelli S C et al. 2014 *Angew. Chem. Int. Ed.* **52** 50 13107. b) Forcén-Vázquez E 2014 Metal Citrate Cubanes: Synthesis, Characterization and Properties. Prensas de la Universidad, Zaragoza. Chapters 6 and 7. https://zaguan.unizar.es/record/30653?ln=es
[7] Ludueña G A et al. 2011 *Chem. Mater.* **23** 1424.

Acknowledgements

Funding was provided by the Ministerio de Ciencia e Innovación (Spain) under Grants MAT2011-27233-C02-01 and MAT2011-27233-C02-02 with cofinancing from the European Union Regional Development Fund (FEDER), and under the program CONSOLIDER-INGENIO in Molecular Nanoscience (reference CSD 2007-00010). EFV thanks the Ministry of Education (Spain) for a pre-doctoral scholarship under the program 'Becas y Contratos FPU' (reference AP2009-4211). Funding was also provided by the Diputación General de Aragón (Spain). The authors acknowledge the services of the Servicio General de Apoyo a la Investigación – SAI, University of Zaragoza

Supplementary Material

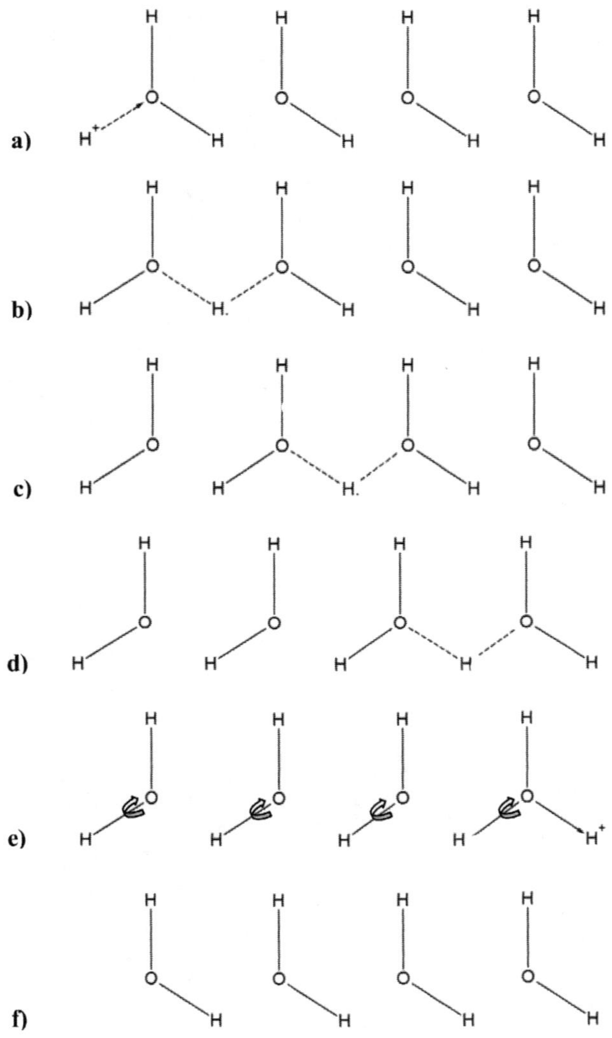

Figure S1. Schematic representation of a proton jump along an H-bonded water chain through the Grotthuss Mechanism. Extracted from reference 6b.

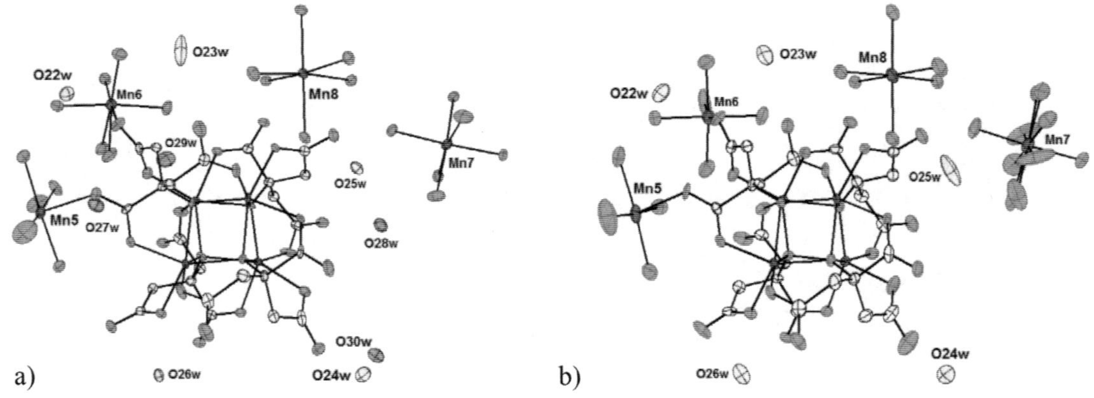

Figure S2. A complete cubane attached to a bridging and a pendant Mn(II) unit and two $[Mn(H_2O)_6]^{2+}$ as counterions and free water molecules for compounds **1** and **1'**. Single crystal X-ray diffraction data. Displacement ellipsoids: 50%

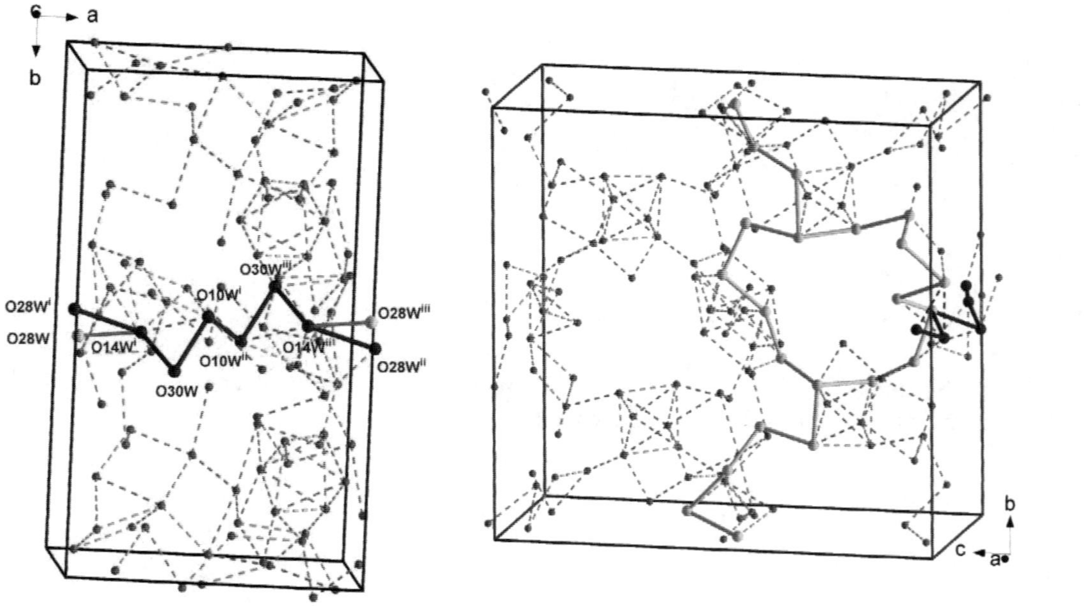

a) b)

Figure S3. Different H-bonded water pathways traversing the unit cell of 1. a) 1D chain and b) cyclic motif. i: -x, 1-y, -z, ii: 1+x, y, z, iii: 1-x, 1-y, -z.

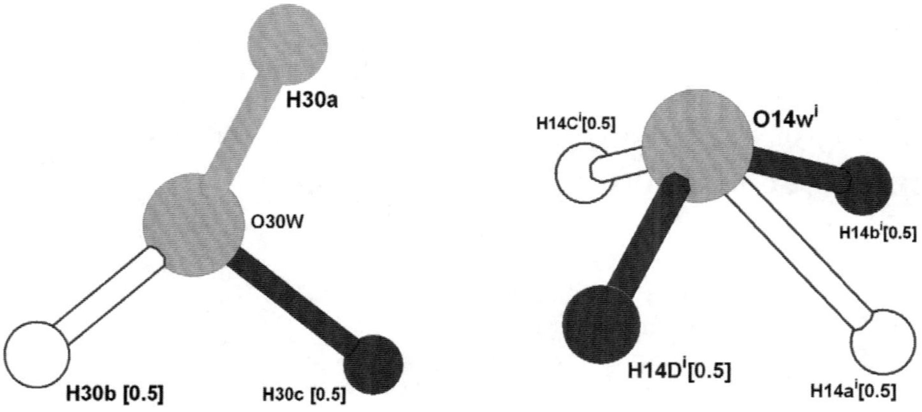

a) **b)**

Figure S4. Experimental disorder found around water oxygen atoms. Occupancies are in brackets. i: -x, 1-y, -z.

Table S1. Distances and angles between the atomic sites involved in the mechanisms detailed in the text. Symmetry operations: i: -x, 1-y, -z; ii: 1+x, y, z; iii: 1-x, 1-y, -z.

Atom A ... Atom B	Distance (Å)
H14d ... O30w	1.93 (3)
H14d ... H30b	1.39 (4)
H30c ... O10w^{i}	1.74 (4)
H30c ... H10b	0.73 (3)
H10c^{i} ... O10w^{ii}	1.88 (2)
H10c^{i} ... H10c^{ii}	0.94 (4)
H14c^{ii} ... O28w^{iii}	1.94 (3)
H28a^{iii} ... O14w^{iii}	2.18 (3)
H28a^{iii} ... H14c^{iii}	1.55 (4)
O28w^{iii} ... O28w^{ii}	1.94 (3)

Neutron Powder Diffraction study of the Magnetic Ionic Liquid Emim[$FeCl_4$] and its deuterated phase

A. García-Saiz[1], I. de Pedro[*1], L. Fernández Barquín[1], M.T. Fernández-Díaz[2] J. A. Blanco[3] and J. Rodríguez Fernández[1]

[1] CITIMAC, Facultad de Ciencias, Universidad de Cantabria, Santander 39005, Spain.
[2] Institut Laue-Langevin, BP 156X, F-38042 Grenoble Cedex, France
[3] Departamento de Física, Universidad de Oviedo, 33007 Oviedo, Spain

E-mail: depedrovm@unican.es

Abstract. A magnetic ionic liquid comprising 1-ethyl-3 methylimidazolium (Emim) cations and tetraclhoroferrate(III) ($FeCl_4$) anions and its deuterated phase were synthetized and characterized magnetically. In both materials, the low temperature dependence of the magnetic susceptibility presents a maximum (around 4 K) related to an antiferromagnetic ordering, but the ordering temperatures are slightly shifted and the curves display different shapes. In addition, the magnetization of the deuterated phase tends to saturate at higher values than that corresponding to the non-deuterated analogue. A comparison of the neutron diffraction patterns above and below the magnetic transition clearly shows that the crystal and magnetic structures of these materials are different. Therefore, the present findings clearly prove that the magnetic exchange interactions that induce three-dimensional magnetic ordering are modified after the deuteration process.

1. Introduction

Magnetic ionic liquids (MILs) have received considerable attention among ionic liquids (ILs). The interest is fuelled by the possibility of tuning the materials properties by means of external magnetic fields or using them in the liquid state to produce nanoparticle-free magnetic emulsions and microemulsions [1]. Moreover, MILs can combine magnetic and ILs properties with additional intrinsic thermochromic, magneto-electrochromic or luminescent properties depending on the enclosed paramagnetic ion used [2]. The combinations of different rare-earth (europium, neodymium, dysprosium, etc.) chiral amionacids, bimagnetic ions or heteroanions have been studied for applications such as transport and separation of materials, separation of greenhouse gases (CO_2) through supported magnetic ionic liquids membranes, magnetic surfactants or esterification of oleic acid to biodiesel [3].

MILs have also recently received attention from the experimental and theoretical point of view concerning to non-covalent interactions and cohesion properties [4]. Many groups have focused their studies on the dipole-dipole interactions, and hydrogen bonds within the crystal and molecular structures in order to improve the understanding of these forces, with the aim of tailoring their potential technological applications [5]. Importantly, in MILs with paramagnetic ions, such as those presented here, it is also necessary to investigate other non-covalent interactions, like halogen-halogen

Journal of Physics: Conference Series **663** (2015) 012008 doi:10.1088/1742-6596/663/1/012008

(between the nearest metal complex anions) or anion-π (between the anion and cation) as these forces play an important role in the organization of the structural units [6]. In addition, when these interactions are strong enough, interesting collective electrical and magnetic phenomena can arise, such as ferroelectricity in the bis(imidazolium) pentachloroantimonate (III), $(C_3N_2H_5)_2SbCl_5$ [7], or a three-dimensional (3D) magnetic ordering in MILs based on tetrahaloferrate ions [8, 9]. In this last family, recent research shows that a smaller chain length in the organic cation enhances the efficiency in the transmission of the magnetic interactions resulting in an increase in the Néel temperature, T_N (from $[Bmim]^+$ that shows no 3D magnetic ordering through $[Emim]^+$, with T_N = 3.8 K [10], to $[Dimim]^+$, with T_N = 5.6 K [11]) [Bmim: 1-butyl-3-methylimidazolium and Dimim: 1,3-dimethylimidazolium]. Moreover, the change of external conditions, like pressure, has a strong influence in the magnetic coupling of Emim[$FeCl_4$], varying from antiferromagnetic (AF) to ferrimagnetic ordering [12]. In this contribution we focus on ascertaining the nature of the magnetic structure of this last MIL, using neutron diffraction experiments. The presence of H atoms in the cation part (H= 11) of Emim[$FeCl_4$] predicted low statistics in the neutron diffraction pattern. Therefore, in order to avoid the large incoherent component of the hydrogen, which creates a large continuous background and complicates the structural resolution, we have also prepared the fully deuterated sample, Emim[$FeCl_4$] d11. Herein, we present the comparative study considering magnetic susceptibility and neutron powder diffraction data.

2. Results and discussion

The zero-field-cooled (ZFC) molar susceptibility (χ_m) at magnetic field of 1 kOe of Emim[$FeCl_4$] d11 displays a paramagnetic behaviour over most of the temperature range studied, with an effective paramagnetic moment, μ_{eff} = 5.70(1) μ_B/Fe ion, and a paramagnetic Curie temperature, θ_p = -3.0(1) K. These data anticipate an overall antiferromagnetic behaviour in good agreement with their non-deuterated phase (μ_{eff} = 5.66(1) μ_B/Fe ion and and θ_p = -2.5(1) K) [8, 9]. At low temperatures (see Figure 1), χ_m of Emim[$FeCl_4$] d11 increases and reaches a maximum at approximately 4.0 K, which suggests the existence of a long-range magnetic ordering. The most striking features of these curves being a small displacement of ordering temperature and the different shape with respect to the non-deuterated phase. These facts indicate that the magnetic coupling is modified with the deuteration process. Moreover, the magnetization of Emim[$FeCl_4$] d11 tends to saturate above 40 kOe, with higher values than those corresponding to the non-deuterated analogue (M ≈ 4.9 μ_B/Fe ion), and it is near the expected fully-saturated value of 5 μ_B/Fe for Fe (III) ion (see Figure 2). As occurs in the non-deuterated compound, the magnetization shows no-hysteresis, thereby discarding the existence of any ferromagnetic component.

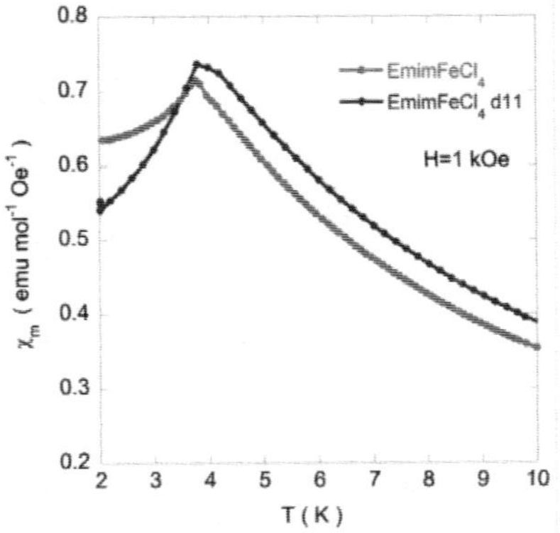

Figure 1. Low temperature ZFC magnetic susceptibility for Emin[$FeCl_4$] and its deuterated phase.

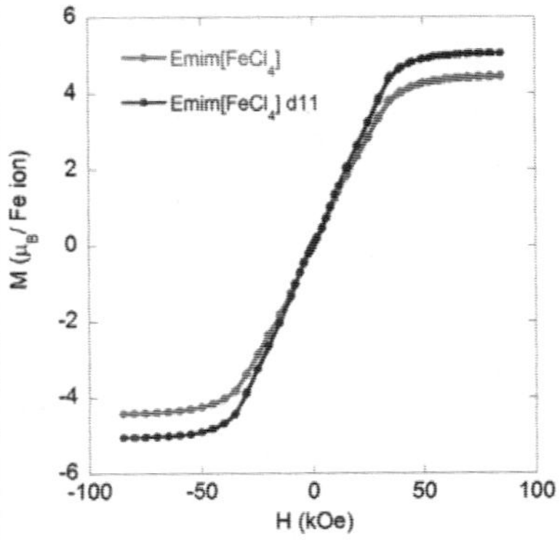

Figure 2. Magnetization of Emin[$FeCl_4$] and its deuterated phase measured at 2 K.

Neutron powder diffraction patterns of Emim[$FeCl_4$], from 270 to 2 K, and of deuterated analogue at 10 and 2 K were collected on D1B diffractometer at the ILL (λ = 2.52 Å). In these experiments, about 2 g of the sample mass was frozen and then milled before it placing in a cylindrical vanadium container pre-cooled at N_2 liquid. The samples were rapidly placed in the cryostat at temperature below the melting point (285K), checking that no preferential orientation was present. The diffraction patterns were collected in the Bragg angular range $5 \leq 2\theta \leq 70^o$. The 2D contour plots of the temperature dependence of the neutron diffraction data of the non-deuterated compound is shown in Figure 3). On one hand, a structural phase transition below 210 K is detected in the neutron diffraction patterns, which seems to be stable down to 2 K. This feature could be related to dynamical reorientations of the cations or anions during the cooling process. Therefore, the appearance of a new peak at lower angle regions at 14.5° (d = 10.06 Å) and the splitting of some crystallographic peaks (such as the one located at $2\theta \approx 45°$) could be indicative of a loss of symmetry. This issue differs from the data reported for the Dimim[$FeCl_4$] [11], in which the structural phase transition changes the symmetry from $P2_1$ (R.T) to $P2_12_12_1$ at 285 K. Additionally, the presence of extra magnetic signals below 4 K reveals the occurrence of a long-range magnetic ordering compatible with the magnetization results.

The high temperature crystal structure was refined starting from the previously reported single-crystal structure model [13]. The best Rietveld refinement from powder diffraction data at 270 K a framework, which projected along *b* axis, can be described as a stacking of $[Edimim]^+$ and $[FeCl_4]^-$ intercalated layers, with Fe···Fe distances larger than 6.5 Å inside the layer [monoclinic space group $P2_1/c$; a = 9.424(3) Å, b = 14.662(4) Å, c = 12.449(9) Å, β = 129.89(2)°, V= 1319.2(4) Å^3]. The anion groups $[FeCl_4]^-$, slightly distorted, form subnets in the *a,c* plane with a change of orientation from layer to layer and a displacement of the subnet with respect to the next one. The cations, $[Emim]^+$, also form subnets in the *a,c* plane but are stacked nearly identically above one another, lying parallel to each other along *b* direction (see Figure 4) in good agreement with single-crystal structure model.

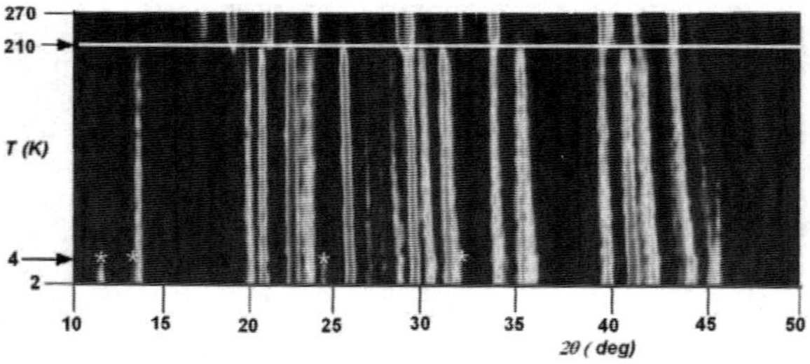

Figure 3. 2D contours of the temperature dependence of the D1B patterns between 270 and 2 K for [Emim][$FeCl_4$]. The (*) dots show the magnetic contributions. The line at 210 K shows the change of the crystalline structure.

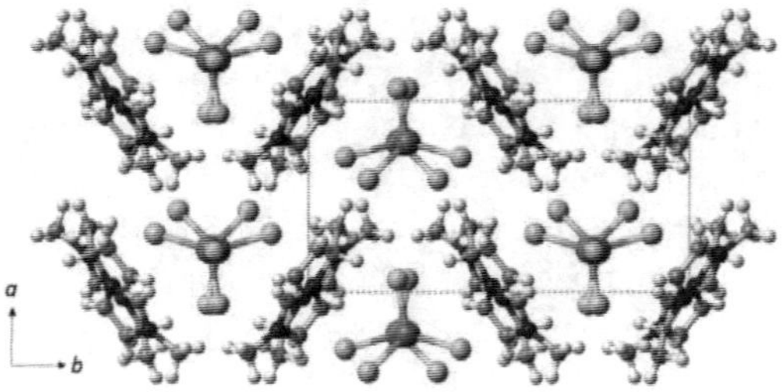

Figure 4. Crystal packing in the *a,b* plane of Emimim[$FeCl_4$] crystal structure at 270 K. Red (iron), green (chloride), black (carbon), blue (nitrogen) and white (hydrogen). The blue dashed square represent the unit cell.

A comparison of the D1B patterns for Emin[$FeCl_4$] (red) and its deuterated phase (blue) at 2 K is plotted in Figure 5. The data clearly show differences in the structural and magnetic peaks. Therefore, the intermolecular interactions between anions and cations (ionic, hydrogen bonding and van der Waals forces) are modified with the deuteration process. These forces affect the local- and long-range molecular ordering, which subsequently modify the magnetic couplings between an iron ion and its first shell of neighbouring iron ions changing the superexchange anion-anion interactions [14]. In order to try to resolve the low temperature crystal structure of both MILs, high-resolution neutron diffraction data at 10 K on D2B (λ = 1.594 Å) were collected (see Figure 6). A comparative view of the patterns clearly shows a better signal/ratio in the deuterated phase and confirms the different low temperature crystal structures.

Only for the deuterated compound was found a satisfactory monoclinic crystal structure [M20 =16.8; F20 = 10.6] using the WINPLOTR programme [15], with unit cell parameters of a = 9.560 (1) Å, b =13.946 (1) Å, c = 14.460 (1) Å, β = 97.13(1). The analysis was consistent with the $P2_1/m$ space group. The experimental, calculated and difference diffraction profiles are displayed in Figures 5 and 6 and the resolution of the crystal structure is now in progress. Focusing on the long-range magnetic ordering of the deuterated phase, below T_N, the additional magnetic peaks can be indexed with the propagation vector k = (0.5, 0, 0.5) (Figure 5), indicating that the magnetic unit cells is twice larger than the crystallographic one in a and c directions. This approach allows us to see some important differences when compared to Dimim[$FeBr_4$] analogous [6], in which the magnetic and nuclear unit cells are the same.

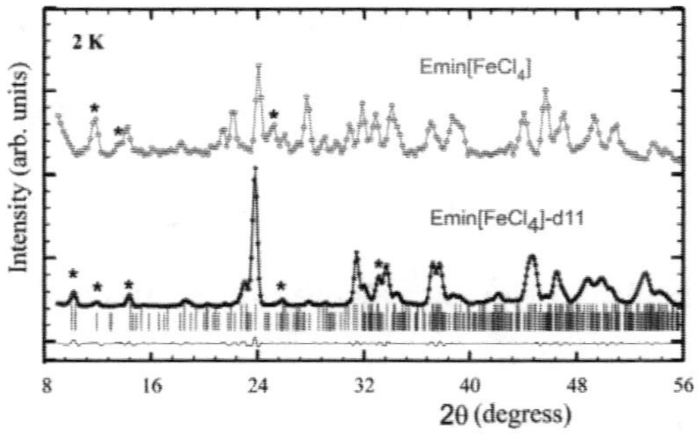

Figure 5. Neutron pattern for Emim[$FeCl_4$] and neutron diffraction profiles for the deuterated phase at 2 K obtained in D1B (λ = 2.52 Å). Positions of the magnetic (second row) profile are also presented. The (*) dots show the magnetic bragg peaks. The curves are shifted for clarity.

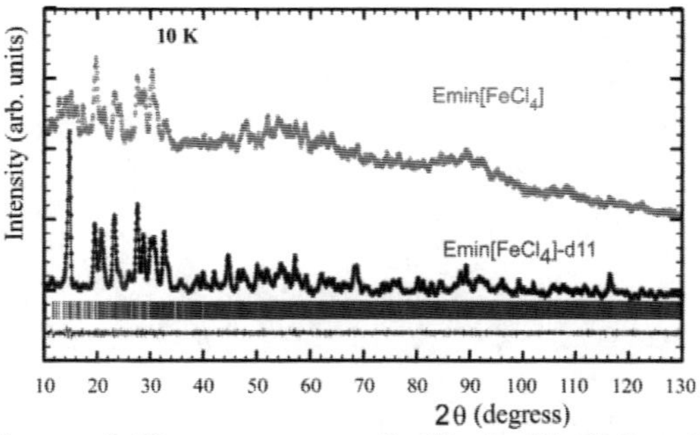

Figure 6. Neutron pattern for Emim[$FeCl_4$] and neutron diffraction profiles for the deuterated phase at 10 K obtained in D2B(λ = 1.594 Å). The curves are shifted for clarity.

3. Conclusions

Neutron data evidence a solid-solid transition in Emim[$FeCl_4$] near 210 K, this crystal structure being stable down to 2 K. The appearance of extra magnetic peaks below 4K in neutron diffraction data confirms the onset of a three-dimensional antiferromagnetic ordering. Finally, we have identified via magnetic and powder diffraction studies that these MILs change their magnetic behaviour with the deuteration process. The 3D antiferromagnetic ordering detected by magnetic measurements must take place via super-exchange coupling, the Fe–Cl···Cl–Fe interaction being the main exchange pathway. In addition, indirect exchange coupling of the type Fe-Cl···Im···Cl-Fe (Im = Imidazolium) could be present in both compounds. Taking in account that the strength of the magnetic exchange pathways via these super-exchange couplings depend on the following paramenters: (i) Cl ··· Cl distance, (ii) Fe– Cl ··· Cl –Fe angles and (iii) the Fe– Cl ··· Cl –Fe torsion angle (τ), these distances and angles should be quite similar, as inferred from the magnetic susceptibility measurements, being different their crystal structures. The present findings illustrates very well how small bonding-changes can have a profound effect on both crystal and magnetic structures of these materials whose framework is linked by non-covalent interactions.

4. Acknowledgements

Part of this work was carried out at Institute Laue-Langevin (ILL) of Grenoble and was partially financially supported by MEC research project MAT2011-27573-C04.

5. References

[1] Okuhata M, Funasako Y, Takahashi K and Mochida T 2013 *Chem. Commun.* **49** 7662.
[2] Torimoto T, Tsuda T, Okazaki K and Kuwabata S 2010 *Adv. Mater.* **22** 1196.
[3] Santos J, Albo A and Irabien R 2014 *Soc. Chem. Adv.* **4** 40008.
[4] Steed, J W and Atwood J L *Supramolecular chemistry*. John Wiley & Sons: 2009.
[5] Dupont, J 2004 *J. Braz. Chem. Soc* **15** 341.
[6] García-Saiz A, de Pedro I, Migowski P, Vallcorba O, Junquera J, Blanco J A and Rodríguez Fernández, J 2014 *Inorg. Chem.* **53** 8384.
[7] Piecha-Bisiorek A, Jakubas R, Medycki W, Florek-Wojciechowska M, Wojciechowski M and Kruk D 2014. *J. Phys. Chem. A* **118** 3564.
[8] de Pedro I, Rojas D P, Albo J, Luis P, Irabien A, Blanco J A and Rodríguez Fernández, J 2010 *J. Phys.: Condens. Matter* **22(29)** 296006.
[9] de Pedro, I, Rojas D. P, Blanco J A and Rodríguez Fernández, J 2011 *J. Magn. Magn. Mater.* **323** 1254.
[10] Yoshida Y, Otsuka A, Saito G, Natsume S, Nishibori E, Takata M, Sakata M, Takahashi M and Yoko T 2005 *Bull. Chem. Soc. Jpn.* **78**, 1921.
[11] García-Saiz A, Migowski P, Vallcorba O, Junquera J, Blanco J. A, González J. A, Fernández-Díaz M. T, Rius J, Dupont J, Rodríguez Fernández J and de Pedro I 2014 *Chem. Eur. J.* **20** 72.
[12] García-Saiz A, de Pedro I, Blanco J A González J A and Rodríguez Fernández, J 2013 *J. Phys. Chem. B* **117** 3198.
[13] Bäcker T, Breunig O, Valldor M, Merz K, Vasylyeva V and Mudring A. V 2011 *Cryst. Growth & Design* **11(6)** 2564.
[14] García Saiz A, de Pedro I, Vallcorba O, Migowski P, Campo I, Barquin, L F, Abrahams I, Motevalli M, Dupont J, Gonzalez J A, Rodríguez Fernández, J 2015 *RSC Adv.* **5** 60835
[15] Roisnel T. and Rodriguez-Carvajal J 2000 WINPLOTR: A New Tool for Powder Difracction: Laboratoire Leon Brillouin (CEA-CNRS) Centre d'Etudes de Saclay, Gif sur Yvette Cedex, France.
[16] J. Rodríguez-Carvajal, J 1993 *Physica B* **192** 55 and later un-published versions; the programme is a strongly modified version of that described by Wiles D. B. and Young, R. A. 1981, *J. Appl. Crystallogr.* **14** 149.

Journal of Physics: Conference Series **663** (2015) 012009 doi:10.1088/1742-6596/663/1/012009

FABADA Goes MANTID to Answer an Old Question: How Many Lines Are There?

D Monserrat[1], A Vispa[1], L C Pardo[1], R Tolchenov[2,3], S Mukhopadhyay[4] and F Fernandez-Alonso[2,5]

[1] Grup de Caracterització de Materials, Departament de Física i Enginyeria Nuclear, ETSEIB, Universitat Politècnica de Catalunya, Diagonal 647, E-08028 Barcelona, Catalonia, Spain
[2] ISIS Facility, Rutherford Appleton Laboratory, Chilton, Didcot, Oxfordshire OX11 0QX, United Kingdom
[3] Tessella Ltd., 26 The Quadrant, Abingdon Science Park, Abingdon OX14 3YS, United Kingdom
[4] ISIS Facility, Rutherford Appleton Laboratory, Chilton, Didcot, Oxfordshire OX11 0QX, United Kingdom. Department of Materials, Imperial College London, Exhibition Road, London SW7 2AZ, United Kingdom
[5] Department of Physics and Astronomy, University College London, Gower Street, London, WC1E 6BT, United Kingdom.

E-mail: luis.carlos.pardo@upc.edu

Abstract. Data fitting and model selection lie at the very heart of the scientific method, and these tasks are often tackled via the use of popular (and typically unquestioned) optimization algorithms of limited validity or flexibility, *e.g.*, least-squares methods. To circumvent these limitations, this work introduces the implementation within the MANTID framework of robust inference methods based on the Bayesian fitting algorithm FABADA. The power of this new MANTID tool is illustrated by revisiting an overarching and recurring question in the interpretation of quasielastic neutron scattering measurements, that is, how many spectral components may (or may not) be justified by the available experimental data.

1. Introduction

Data fitting and model selection underpin much of what is done with scientific data, to verify the adequacy of specific mathematical models and to obtain parameters that best describe the underlying physical phenomena. The classical way of performing data fitting involves the minimization of a so-called *cost function* (CF) such as the celebrated χ^2. For a given physical model, popular algorithms such as Levenberg-Marquardt (LM) seek to minimize the CF in order to find the parameter values (and associated uncertainties) that 'best describe' the data [1]. Some of the limitations associated with these classical approaches have been previously discussed by some of the authors [2, 3]. In brief, LM algorithms are successful when the χ^2 is well behaved, *i.e.,* if no local minima are encountered in the search for the global minimum. Moreover, LM is based on the assumption that the minimum is quadratic and that there is no correlation between fit parameters beyond linear couplings. With these restrictions in mind, model selection is then implemented on the basis of the value of the χ^2 divided by a term that penalizes the addition of extra model parameters, leading to a concomitant reduction in the overall number of degrees of freedom. Such is the rationale behind the concept of a reduced-χ^2 (χ^2_{red}) CF, commonly defined

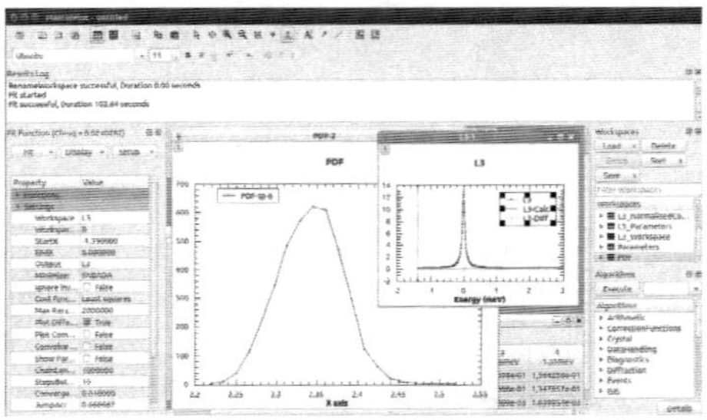

Figure 1. (Color online) Snapshot of MANTID program where the minimizer FABADA has been selected.

as $\chi^2_{red} = \chi^2 \cdot (N - m - 1)^{-1}$, where N is the number of data points and m is the number of fit parameters. Model selection based on χ^2_{red} is faced with three main drawbacks. In addition to the primary assumptions of the LM method discussed above, parameter correlation leads to an effective reduction in the number of real parameters. In the language of statistical analysis, this added complication implies that the CF does not conform to the expected χ^2 Probability Density Function (PDF) [4]. Last but not least, reliance on a single number (the χ^2) is unlikely to represent a general way of performing robust model selection.

Bayesian Inference Methods (BIMs) provide a suitable framework to address the above limitations. First and foremost, BIMs do not rely on specific assumptions about the underlying χ^2 landscape or on putative (and typically unknown) correlations between model parameters. In addition, jumps across local minima are possible [2] and the robustness of the results can always be tested by increasing the (fictive) temperature associated with artificial increments in data uncertainties [5]. Also, the interpretation and visualization of BIM results differ from the more widespread *frequentist* point of view in a number of fundamental ways. In BIMs, PDFs are obtained for each model parameter, as well as for the χ^2. Likewise, model-selection protocols can be implemented in a systematic manner via direct inspection of the χ^2 PDF. This approach can immediately establish the mono- or multi-modal character of this function, and the position of its most-probable value provides a suitable figure of merit for quantitative model selection. Moreover, analysis of the PDFs makes it possible to *assess the robustness* of any criteria we might have chosen to implement model selection. For example, if χ^2 PDFs corresponding to two different models display some overlap in χ^2 space, this result immediately implies that there exists a combination of parameters for one model which, given the available experimental data, is necessarily as likely as the corresponding parameter set for the second (alternative) model.

In the present work, we introduce the first implementation of BIMs within the MANTID framework [6, 7]. As a software package, MANTID is primarily aimed at the visualization and analysis of neutron and muon data using modern software-engineering paradigms, as recently illustrated in Refs. [8, 9, 10]. The added value of this new MANTID tool is illustrated by considering its use in QuasiElastic Neutron Scattering (QENS) data analysis, superseding previous BIMs [11, 12, 13]. The implementation has been done in the release 3.4 and has been added as a new minimizer in the main menu of MANTID as it can be seen in figure 1.

2. A Closer Look Under the Hood: The FABADA Algorithm

Before describing the implementation of BIMs in MANTID, we provide a summary of the underlying Fitting Algorithm for Bayesian Analysis of DAta, hereafter FABADA [3]. Bayes' theorem relates the conditional probabilities between two events [4] and therefore provides the starting point to calculate the likelihood that a given hypothesis is supported by the experimental data. This task may be accomplished by taking proper account of our prior state of knowledge concerning the starting hypothesis as well as the likelihood that the experimental data are described by the proposed hypothesis. In what follows, we assume no previous knowledge about the hypothesis, a situation that amounts to assigning a constant probability to the prior.

Typically, neutron-scattering experiments rely on the acquisition of a given number of neutron counts per channel. When count levels are high, the underlying Poisson distribution describing experimental uncertainties becomes Gaussian, and the probability that a given model is supported by the experimental data becomes proportional to exp $\left(-\frac{\chi^2\{P_i\}}{2}\right)$. In this expression χ^2 is defined as usual, that is, as the quadratic sum of the deviation between data and model weighted by uncertainties in the former (for more details, see Refs. [2, 3]). As expected, the value of χ^2 depends on the specific values of the model parameters. This condition can be made explicit by writing $\chi^2\{P_i\}$ where P_i is used to denote the set of model parameters. FABADA, or any other Markov Chain Monte Carlo (MCMC) method, changes one of the parameters in P_i in a random manner, and accepts the change if the CF (χ^2 in our case, or the energy in Monte Carlo simulations) diminishes. Since there are errors associated with the data (or thermal agitation in the case of a Monte Carlo simulation), there is also a non-zero probability to accept changes in P_i that nonetheless lead to an increase of the CF. In the case of data fitting, the probability of accepting an increase in the CF can be written as

$$\frac{P[H(P_i^{new})|D]}{P[H(P_i^{old}]|D)} = exp\left(-\frac{\chi^2_{new} - \chi^2_{old}}{2}\right) \tag{1}$$

where $H(P_i^{new})$ denotes the hypothesis (model) described by the parameter set P_i^{new}, D represents the experimental data, and $P[H|D]$ denotes the conditional probability that hypothesis H holds given D. This additional allowance for otherwise "illegal" moves that increase the CF enables jumps over small barriers in the χ^2 landscape. Higher barriers can also be avoided by multiplying data uncertainties by a fictive temperature, as detailed in Refs. [2, 5].

The main limitation of the above algorithm is related to the extent one needs to change parameter values so that the optimization process becomes equally sensitive to all of them. To circumvent this difficulty, FABADA uses an adaptive MCMC scheme that avoids unwanted bias to specific parameters. This method is best understood by considering a simple example. Let us assume that a given parameter is changed by an amount Δ_i following the prescription $P_i^{new} = P_i^{old} + \rho\Delta_i$ where ρ is a random number. To make all parameters relevant during the fitting procedure, FABADA changes the value of the step Δ_i in an adaptive manner based on the acceptance ratio. Parameters characterized by a low acceptance ratio would then be given large values of Δ_i so that changes in χ^2 become large. This condition leads to large values of $\chi^2_{new} - \chi^2_{old}$ in Eq. (1) and low acceptance probabilities. The opposite happens to parameters with a high acceptance ratio. Convergence implies that all parameters must be accepted with equal probability, a condition that generally requires different values of Δ_i. The specific values of Δ_i are also related to the curvature of the (multidimensional) χ^2 PDF along the direction of a given parameter and its associated error. A small value of Δ_i implies a strong curvature and a small error.

3. Bayesian Model Selection: How Many Lines Are There?

Falsifiability in science [14] usually operates via model selection. For this reason, understanding the conceptual underpinnings of this process is an important exercise deserving careful thought and consideration. One of the most widely used model-selection criteria is predicated upon the notion that a given model is better than others because it provides a "better fit" to the available data. This condition is only true when the number of parameters across models is the *same* and all parameters remain *uncorrelated*, as correlation diminishes their effective number. In the

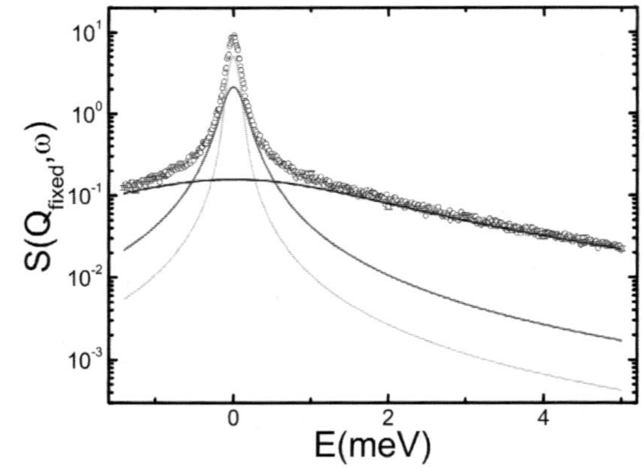

Figure 2. (Color online) Synthetic QENS spectrum to test model selection (black circles). Note the Log-Lin representation of these data. Green, red, and blue lines have been used to depict three L modes with spectral HWHMs 0.02, 0.2, and 1 meV, respectively.

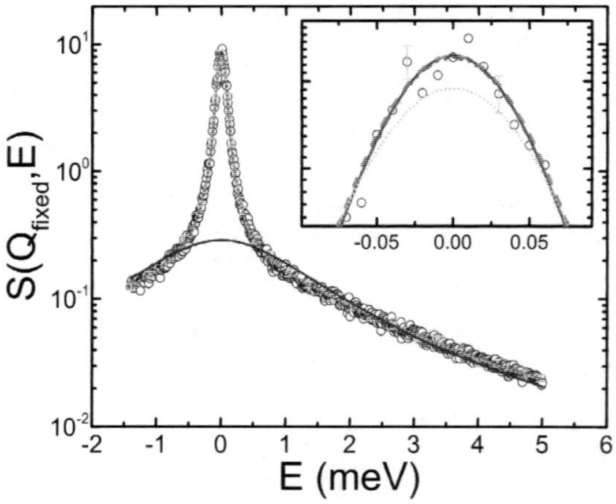

Figure 3. (Color online) Fits to the synthetic QENS data (black circles) using an increasing number of L modes: L1 (solid blue line), L2 (dotted green), L3 (dashed-dot red), L4 (dashed orange), and L5 (solid purple). Note the logarithmic (linear) scales on the main figure (inset).

(metaphorical) words of von Neumann: 'With four parameters I can fit an elephant, and with five I can make him wiggle his trunk' [15]. A model with more parameters usually gives a better fit to the data, and, thus, we must be careful to assess whether the data *and their associated errors* support the new model.

In the context of neutron data analysis, we provide below an alternative answer to an old question posed by Sivia and Carlile in their seminal study *How Many Lines Are There?* [16]. The advantage of the method presented in this work is its flexibility since any parametric model, analytical or not, can be used. We briefly recall that QENS data analysis is oftentimes complicated by the fact that a number of (*a priori* unknown) dynamical processes are all centered around the elastic response of the material. In the limit of very different timescales associated with these processes, visual inspection of the data can be a convenient starting point to construct a suitable prior. The situation becomes less clear when spectral widths become comparable to each other, the extreme case being the occurrence of accidental degeneracies which necessarily lead to a reduction of the dimensionality of our available parameter space. To address this question in a systematic manner, we have generated synthetic QENS data as the linear addition of three equally weighted Lorentzian (L) modes convolved by a Gaussian resolution function with a Half-Width-Half-Maximum (HWHM) of 0.005 meV. The three L modes have been generated with spectral HWHMs of 0.02, 0.2, and 1 meV. To emulate realistic experimental conditions, a Gaussian-distributed relative error of 6% has been added to the QENS spectrum after convolution. These data are shown in Fig. 2.

All data fits have been performed using the current implementation of FABADA in MANTID [7]. To explore model selection with this new MANTID tool, Fig. 3 shows best-fit results for a total of five different models described by an increasing number of L modes, hereafter denoted as Ln where n is the number of L modes. Visual inspection of these results makes it clear that L1 is not capable of describing the data satisfactorily. Beyond this case, a cursory 'χ-by-eye' analysis of these data is no longer possible. A much closer look at specific regions of the spectral response might still allow us to discover that L3 is slightly better than

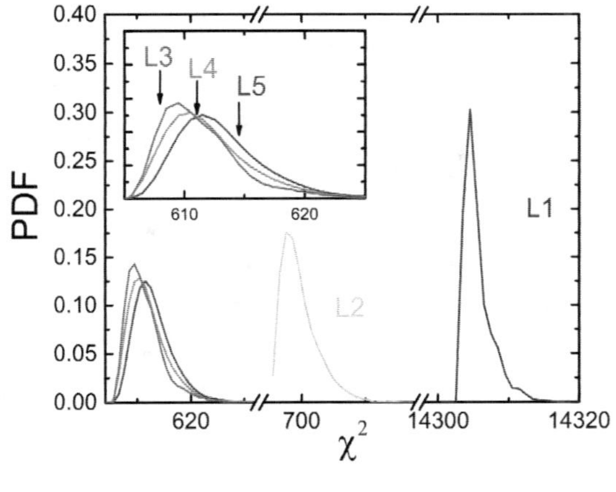

Figure 4. (Color online) χ^2 PDFs for the L1-L5 models shown in Fig. 3. The inset shows an enlargement around L3-L5, showing that L3 is favored over the other two. This figure uses the same color coding as in Fig. 3.

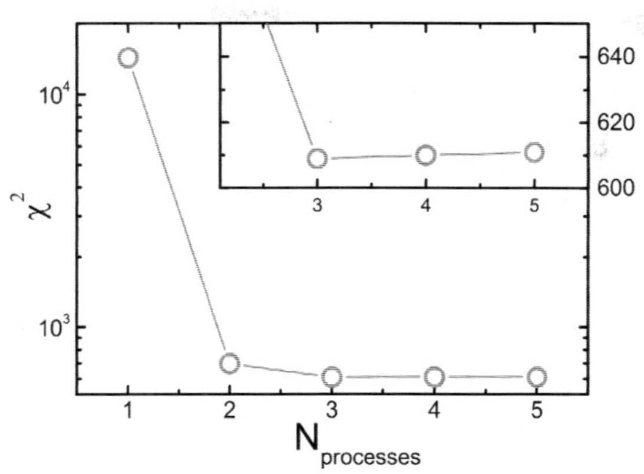

Figure 5. Most-probable values of the χ^2 PDF for models L1-L5. Note the Log(Lin) scale in the main panel (inset).

L2 (*cf.* inset in Fig. 3). For L3, L4, and L5, the quality of the fits remains virtually unchanged. Thus, in this particular case, even a simple model-selection protocol using a χ^2_{red} criterion would favor L3 over more complex descriptions of the experimental data.

To put these considerations on firmer ground, Fig. 4 shows the χ^2 PDFs associated with each Ln model. As it became evident from a visual inspection of Fig. 4, the most probable value of the χ^2 PDF for L1 peaks at $\chi^2 \sim 14305$, more than one order of magnitude above other models. For L3, L4, and L5, we see that these three PDFs share the same onset corresponding to the lowest χ^2 (best fit), as expected from the results shown in Fig. 3. However, the *maximum* of the PDF shifts to higher χ^2 values in going from L3 to L5. This trend is a consequence of an increase in the number of parameters, leading to an overall broadening and concomitant upward shift of the corresponding χ^2 PDF. Thus, *the addition of more parameters beyond L3 is clearly not justified by the experimental data.* From a physical viewpoint, the most interesting situation relates to the differences between L2 and L3. Fig. 4 provides a quantitative answer to this question. As shown in this figure, the PDF for L2 peaks at $\chi^2 \sim 700$ and does not overlap with the corresponding one for L3. In probabilistic terms, the immediate consequence of this result is that *no combination of L2 parameters can provide a better fit than L3.* Therefore, *L3 is favored over L2*, a result that we have established in a robust and quantitative manner.

Although direct inspection of χ^2 PDFs represents a safer way of performing model selection, it might also be desirable to have a simple measure of fit quality. The results presented above make it clear that the most-probable value of the χ^2 PDF could be used to this end. Fig. 4 also brings to the fore a common misconception in model selection, as the minimum χ^2 values associated with a particular model (the choice in popular minimization algorithms) is characterized by a negligibly small probability. To close this discussion, Fig. 5 shows these most-probable values as a function of the number of L modes. As one would expect on the basis of our analysis, its value decreases until it reaches a minimum for L3, followed by a slight increase thereafter due to the broadening of the underlying PDF (see inset in Fig. 4). The attentive reader might also have realized that this procedure is tantamount to the implementation of *Ockham's Razor on our QENS data using models L1-L5.*

Journal of Physics: Conference Series **663** (2015) 012009 doi:10.1088/1742-6596/663/1/012009

4. Conclusions & Outlook

Using the Bayesian fitting algorithm FABADA, we have introduced the implementation of BIMs in the MANTID framework. This new tool enables robust data fitting and model selection well beyond more popular and widespread data-analysis methodologies. To illustrate its use, we have revisited an old (yet still topical) question relating to the number of spectral modes present in QENS data. From this exercise, we highlight the merits associated with the use of the *entire* PDF of the CF instead of relying on single-valued criteria such as the minimum χ^2. This approach allows us to ascertain in a quantitative manner whether any combination of parameters for a given model could provide a better description of the experimental data relative to other models. Although it is true that in this simple case a classical model selection would have performed quite good, we would like to point out that the presented method is able to deal with some cases where classical fitting algorithms would not be able to provide a right answer.

We would like to emphasize that this method does not rely on any assumptions relating to the shape of the underlying χ^2 landscape or possible correlations between fit parameters. For the QENS data explored herein, we have only assumed that the data arises from a counting experiment with sufficiently high count rates such that the underlying Poisson distribution becomes Gaussian. Relaxing this assumption has already been implemented with success [3] and could find further use in the analysis of neutron experiments suffering from inherently low signal levels. Moreover our approach is able to include prior knowledge to the fitting, for example coming from previous determinations as shown in the work of Henao et al. [5]. Classical methods based on χ^2 are not able to deal with non-Gaussian distributed parameters, small number of counts or including previous information.

This work was supported by the Spanish MINECO (grants No. FIS2011-24439 and FIS2012-39443-C02-01) and by the Government of Catalonia (grants No. 2009SGR-1003 and 2009SGR-0581). F.F.-A. gratefully acknowledges financial support from the UK Science and Technology Facilities Council.

References

[1] Acton F S 1990 *Numerical Methods that (Usually) Work* (The Mathematical Association of America)

[2] Pardo L C, Rovira-Esteva M, Busch S, Moulin J F and Tamarit J Ll 2011 *Phys. Rev.* E **84** 046711

[3] Pardo L C, Rovira-Esteva M, Busch S, Ruiz-Martín M D and Tamarit J Ll 2011 *J. Phys.: Conf. Ser.* **325** 012006

[4] Sivia D S and Skilling J 2006 *Data Analysis: A Bayesian Tutorial* (Oxford University, New York)

[5] Henao A, Rovira-Esteva M, Vispa A, Tamarit J Ll, Guardia E and Pardo L C 2013 *J. Phys.: Condens. Matter* **25** 454217

[6] Arnold O *et al* 2014 *Nucl. Instrum. Meth.* A **764** 156

[7] The latest version of the MANTID software package can be downloaded free of charge from dx.doi.org/10.5286/SOFTWARE/MANTID

[8] Jackson S 2014 *Rutherford Appleton Laboratory Technical Report RAL-TR-2014-010* (Didcot, UK). Url: epubs.stfc.ac.uk/work/12240844

[9] Mukhopadhyay S 2014 *Rutherford Appleton Laboratory Technical Report RAL-TR-2014-005* (Didcot, UK). Url: epubs.stfc.ac.uk/work/12135206

[10] Jackson S, Krzystyniak M, Seel A G, Gigg M, Richards S E and Fernandez-Alonso F 2014 *J. Phys.: Conf. Ser.* **571** 012009

[11] Howells W S, Garcia Sakai V, Demmel F, Telling M T F and Fernandez-Alonso F 2010 *Rutherford Appleton Laboratory Technical Report RAL-TR-2010-006* (Didcot, UK). Url: epubs.stfc.ac.uk/work/52040

[12] Fernandez-Alonso F, Bermejo F J, McLain S E, Turner J F C, Molaison J J and Herwig K W 2007 *Phys. Rev. Lett.* **98** 077801

[13] Fernandez-Alonso F, McLain S E, Taylor J W, Bermejo F J, Bustinduy I, Ruiz-Martin M D and Turner J F C 2007 *J. Chem. Phys.* **126** 234509

[14] Popper K R 2002 *Conjectures and Refutations: The Growth of Scientific Knowledge* (Routledge Classics)

[15] Dyson F 2004 *Nature* **427** 297

[16] Sivia D S and Carlile C J 1992 *J. Chem. Phys.* **96** 170

Journal of Physics: Conference Series **663** (2015) 012010 doi:10.1088/1742-6596/663/1/012010

Low–energy spin–wave excitations in amplitude–modulated magnetic structure of $PrNi_2Si_2$

J. A. Blanco[1], **B. Fåk**[2], **J. Jensen**[3] , **M. Rotter**[4], **A. Hiess**[5], **D. Schmitt**[6], **P. Lejay**[7]

[1]Departamento de Física, Facultad de Ciencias, 33007 Universidad de Oviedo, Spain
[2]Institut Laue Langevin, 156X, 38042 Grenoble cedex 9, France
[3]Niels Bohr Institute, Universitesparken 5, DK–2100 Copenhagen, Denmark
[4]Max–Planck Institute for Chemical Physics of Solids, 01187 Dresden, Germany
[5]European Spallation Source ESS AB, P.P. Box 176, SE–22100 Lund, Sweden
[6]ISTerre, CNRS, Université de Grenoble I, F–38041 Grenoble, France
[7]Institut Néel, CNRS, BP 166, 38042 Grenoble Cedex 9, France

E-mail: `jabr@uniovi.es`

Abstract. Inelastic neutron scattering (INS) experiments and random phase approximation calculations have been used to investigate the low–energy spin–wave excitations in $PrNi_2Si_2$. The modulated magnitude of the ordered magnetic moments of Pr^{3+} ions implies that the associate, longitudinally polarized magnetic excitations are more intense and dispersive than the usual transverse spin waves. Within the random phase approximation the results are in good overall agreement with the predictions made by the model determined previously from the paramagnetic excitations. The most unusual observation is the well–defined amplitude mode detected close to the magnetic Bragg point existing simultaneously with the phason mode. At low energies, an extra mode is observed to hybridize with the magnetic phasons in the neighborhood of the magnetic Brillouin zone center. A magnetoelastic interaction between the magnetic excitations and the longitudinal phonons is able to explain part of the disturbances, but it is concluded that the extra mode must be of some other, unknown origin.

1. Introduction

In incommensurate magnetic structures, the only low–energy mode that is expected to be is the pseudo–Goldstone mode of broken translations, known as a phason [1]. This excitation is a purely longitudinal mode associated with the phase of the complex order parameter and represent a modulation of the magnetic moment only. This phason mode is quite unusual, where typically the low–energy collective modes are transverse spin–waves, which correspond to small rotations of the magnetic moments away from their ordered axes. In the spin–wave theory, longitudinal modes are often expected to be highly damped, and then hard to be observed [2]. For a recent notable exception to this rule is the case of amplitude–modulated (AM) magnetic structure of $PrNi_2Si_2$ [3]. It presents a peculiar magnetic character: it is one of the few examples [4, 5, 6, 7, 8, 9, 10] in nature where the magnetic ions, Pr^{3+}, have a longitudinal AM magnetic structure below $T_N = 20$ K, which is stable down to 0 K. In this longitudinally polarized phase, the length of the magnetic moments is modulated and the magnitude of the exchange field changes from site to site. In this way, the material is characterized by an Ising–type magnetic structure, where the magnetocrystalline anisotropy confines the magnetic moments along the c–axis. In most "equal-moment" systems the moments saturate at low temperatures, in which case the longitudinal excitations are quenched, allowing only the presence of transversely polarized spin waves. The AM magnetic structure of $PrNi_2Si_2$ at low temperature (4 K) is well described by a single wave vector $\vec{k} = (0, 0, k_z)$ with $k_z = 0.87$ (in r.l.u.) such that the magnetic moment at the site l is given by the expression $< M_l^z >= \cos(2\pi z k_z)$, where z is the coordinate of the Pr^{3+} ions [11]. The observation of a third harmonic at even lower temperatures reflects a weak tendency to squaring up of the AM moment, but the observed moment ratio between the third and first harmonics is far from the limit of perfect squaring [11, 12]. The dispersion and intensities of both longitudinal and transverse excitations in $PrNi_2Si_2$ were measured along the high–symmetry directions below T_N in several INS experiments [3, 13, 14, 15, 16, 17]. Information about both the experimental details and all the longitudinal and transverse excitations observed in $PrNi_2Si_2$ can be found elsewhere [3]. In this contribution, we now present the results of the dispersive lowest–energy excitations and a discussion of the possible interpretation of this dispersion considering a coupling to the longitudinal phonons.

2. Results and discussion

The excitations observed in the ordered phase relate in a direct way to the excitations in the paramagnetic phase via the level diagram shown in Figure 1 [18]. The *cc* polarized paramagnetic excitations are the cooperative excitations deriving from the transition between the $\Gamma_1^{(1)}$ and Γ_2 states at zero exchange field, and the longitudinal magnetic excitations in the ordered phase derive from the same transition at non-zero exchange fields. Similarly, the $\Gamma_1^{(1)}$–$\Gamma_5^{(1)}$ excitations in the paramagnetic phase become the transverse spin waves in the ordered phase. In a comparison with the paramagnetic phase, the behavior of the excitations in the ordered phase is more complicated because the translational symmetry is reduced, but, on the other hand, they are more well defined since the temperature is lower.

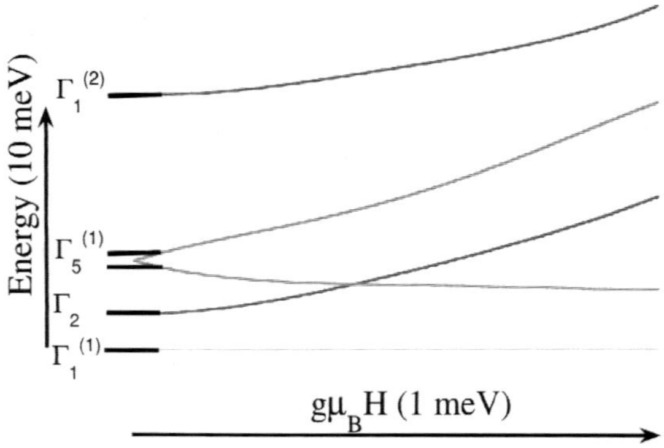

Figure 1. (Color online) The mean field low–lying excitation energies for the Pr^{3+} ions as functions of an exchange field ($g\mu_B H$) along the z axis, calculated with the use of the crystal field parameters given by reference [16]. The classification of the levels is the one applying at zero field.

The INS cross sections measured in the ordered phase of $PrNi_2Si_2$ have been analyzed in terms of the CF and exchange parameters determined from the paramagnetic properties of the system. The previous analysis and the present one are both based on the random phase approximation (see Figure 2). By involving interactions between neighbors 8 layers apart along the c axis, the RKKY exchange interaction in $PrNi_2Si_2$ is long ranged [13]. This indicates that the RPA should be trustworthy, in particular in the zero temperature limit where line–width effects are unimportant. The agreement between experiments and theory is quite good at energies above the range of the phason mode above 2 meV (not shown, see reference [3]). The analysis of the paramagnetic excitations leaves some room for modifications of the exchange parameters and utilizing the remaining degrees of freedom, it might be possible to improve the comparison

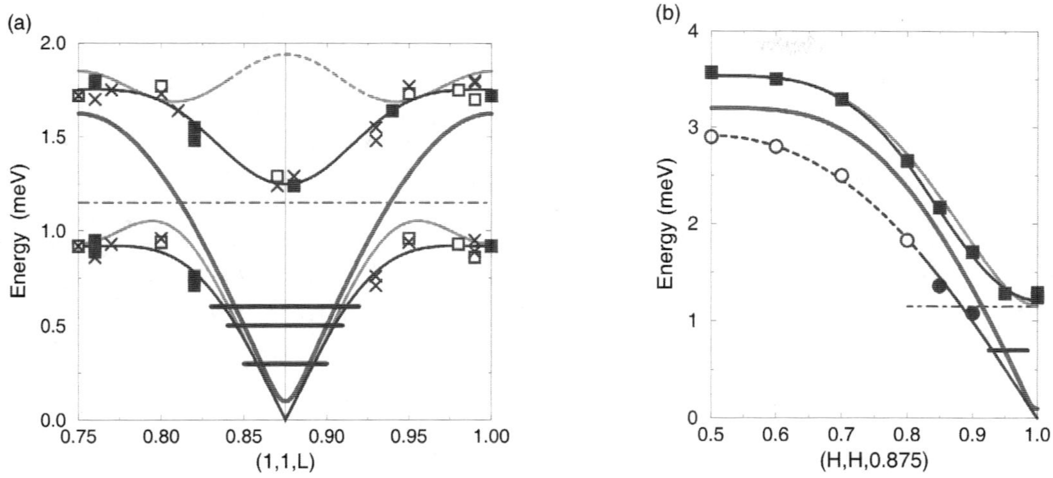

Figure 2. (Color online) The dispersion at low energies of the magnetic excitations in $PrNi_2Si_2$ at $T = 1.6$ K for the wave vectors (a) $(1, 1, L)$ and (b) $(H, H, 0.875)$. The different blue symbols denote the experimental results. The solid blue squares are the constant Q–scan results obtained at the wave vectors shown, the open blue squares in (a) denote results obtained in the neighboring magnetic Brillouin zones shifted to the present one, and the blue crosses show all the experimental results when being reflected with respect to the zone center at $(1, 1, 0.875)$. The circles in (b) denote low intensity peaks, and the open ones indicate peaks that are so weak that their presence may be questioned. In both figures the horizontal line segments indicate results obtained in constant–E scans, and the solid blue lines are least squares Fourier fits to (lines connecting) the experimental points. The solid red lines show the calculated energies of the phason mode. The green lines are the final excitation energies when the longitudinal phonons and the longitudinal magnetic excitations interact according to the magnetoelastic model. The horizontal dot–dashed violet lines indicate the position of a possible flat level interacting with the phasons (see text).

in some of the cases. The most noticeable discrepancy is that the RPA model predicts the energy dispersion of the phason mode along $(H, H, 0.875)$ at 1.6 K to be about 10% smaller than observed [see Figure 2(b)]. This defiance may possibly be removed by a modification of the exchange parameters, but we have not tried to do that, because it might as well be related to the interaction occurring between the phasons and the extra mode at about 1.2 meV.

The major problem in the analysis of the experimental excitation spectrum in $PrNi_2Si_2$ is the clear observation that the phason mode is interacting with an extra level lying at an energy of about 1.2 meV (see Figure 2). It can be excluded that this level should originate from a $4f$ orbital of the Pr^{3+} ions. One possible candidate is the longitudinal acoustic phonons transferred by the modulation of the ordered moments to the relevant range in wave vectors and energies. Using a magnetoelastic model we are able to explain some of the hybridization phenomena, [3])but it failed in an essential way to predict the right behavior in the close neighborhood of the magnetic Brilluoin zone center. The possibility that low–symmetry interactions should have increased the effective period for the longitudinal excitations by a factor of two, and hence introduced an extra low lying amplitude mode, can also be dismissed. The RPA might be wrong in placing the principal amplitude mode at energies about 3 meV instead of 2 meV. The RPA also predicts that the intensity of the amplitude mode should be small relatively to the phason mode at all wave vectors. Since we expect that, at least, this more qualitative prediction should be correct, the large intensity of the upper mode compared to the lower one at the zone boundaries in Fig. 2(a) disqualifies this possibility. The well–defined systematic hybridization behavior observed in the experiments seems to exclude that the extra level should be an experimental artefact. Is it possible that this extra level is established by the electrons on the Ni ions or by the conduction electrons? We do not think so, since, at the low energies in question, these outer electrons are expected just to constitute a passive medium for establishing the RKKY interaction. The only remaining possibility is, as far as we can figure out, that the level is caused by "impurities". The most likely origin of a well–defined impurity level would be that a minor percentage of the Pr ions are displaced from their right position in the crystal. More experimental investigation of the properties of the phonons and the significance of the crystal quality have to be done in order to clarify the cause for the extra level.

3. Conclusions

In summary, we have shown that $PrNi_2Si_2$ is an ideal system for studying the behavior of the magnetic excitations in an amplitude modulated magnetic structure. The most unusual observation being the well–defined amplitude mode observed close to the magnetic Bragg point existing simultaneously with the phason mode. All of them are well accounted for in an RPA model, except for the mysterious hybridization of the phasons with an extra level of unknown origin. This conclusion compares in many ways with the one made in the case of another system with Pr^{3+} ions placed in metallic surroundings, namely that of elemental Pr metal [19, 20].

Acknowledgments

Financial support has been received from Spanish MEC and FEDER Grant No. MAT2011-27573-C04-02 and FC-15-GRUPIN14-037 (Asturias Government, Spain).

References

[1] Koebler U, Hoser A 2009 *Renormalization Group Theory, Springer Series in Materials Science, Heidelberger*
[2] Jensen J and Mackintosh A R 1991 *Rare Earth Magnetism, Clarendon Press, Oxford*
[3] Blanco J A, Fåk B, Jensen J, Rotter M, Hiess A, Schmitt D, Lejay P 2013 *Phys. Rev. B* **87** 104411
[4] Shirane G and Takei W J 1962 *J. Phys. Soc. Jpn.* **17** (BIII) 35
[5] Blanco J A, Gignoux D, Gomez–Sal J C, Rodríguez Fernández J and Schmitt D 1992 *J. Magn. Magn. Mat.* **104–107** 1285
[6] Espeso J I, Soldevilla J G, Blanco J A, Rodríguez Fernández, Gómez–Sal J C, Fernandez-Diaz M T 2000 *Eur. Phys. J. B* **18** 625
[7] Koehler W C 1972 *Magnetic Properties of Rare Earths* Edited by R. J. Elliot, Plenum, New York.
[8] Blanco J A, Gignoux D and Schmitt D 1992 *Z. Phys. B–Condensed Matter* **89** 343
[9] Nicklow R M and Wakabayashi N 1995 *Phys. Rev. B* **57** 12425
[10] Blanco J A, Gignoux D, Gomez–Sal J C, Rodríguez Fernández J and Schmitt D 1992 *J. Magn. Magn. Mat.* **104–107** 1285
[11] Blanco J A, Fåk B, Ressouche E, Grenier B, Rotter M, Schmitt D, Rodríguez–Velamázan J A, Campo J and Lejay P 2010 *Phys. Rev. B* **82** 054414
[12] Rossat–Mignod J 1987 *Methods in Experimental Physics, Ed. by K. Skoeld and D. L. Price, Academic Press, Amsterdam, Vol. 23, p. 131*
[13] Blanco J A, Nicklow R M and Schmitt D 1997 *Phys. Rev. B* **56** 11666
[14] Blanco J A, Fåk B, Nicklow R M, Roessli B and Schmitt D 1997 *Physica B* **234–236** 756
[15] Blanco J A, Gignoux D, Gomez–Sal J C and Schmitt D 1992 *J. Magn. Magn. Mat.* **104–107** 1273
[16] Blanco J A, Schmitt D and Gomez–Sal J C 1992 *J. Magn. Magn. Mat.* **116** 128
[17] Blanco J A, Gignoux D and Schmitt D 1992 *Phys. Rev. B* **45** 2529
[18] Rotter M, Manh D L, Boothroyd A T, Blanco J A 2012 *J. Phys.: Condens. Matter* **24** 213201
[19] Clausen K N, McEwen K A, Jensen J, Mackintosh A R 1994 *Phys. Rev. Lett.* **72** 3104
[20] Clausen K N, Srensen S A, McEwen K A, Jensen J, Mackintosh A R 1995 *J. Magn. Magn. Mat.* **140** 735

Journal of Physics: Conference Series **663** (2015) 012011 doi:10.1088/1742-6596/663/1/012011

Guiding Criteria for Instrument Design at Long-pulse Neutron Sources

J.P. de Vicente[1,2], F. Sordo[1,2], J.M. Perlado[2], F.J. Bermejo[3], and F. Fernandez-Alonso[4,5]

[1] Consorcio ESS-Bilbao, Polígono Ugaldeguren III, Pol. A, 7B. 48170 Zamudio, Spain
[2] Instituto de Fusión Nuclear, José Gutiérrez de Abascal, 2, 28006 Madrid, Spain
[3] Instituto de Estructura de la Materia, IEM-CSIC, Consejo Superior de Investigaciones Científicas, Serrano 123, 28006 Madrid, Spain
[4] ISIS Facility, Rutherford Appleton Laboratory, Chilton, Didcot, Oxfordshire OX11 0QX, United Kingdom
[5] Department of Physics and Astronomy, University College London, Gower Street, London, WC1E 6BT, United Kingdom

E-mail: jpvicente@essbilbao.org

Abstract. We introduce and describe general criteria which characterize long-pulse neutron sources, with a view to guiding and facilitating subsequent instrument design and optimization for specific applications. The ensuing analysis shows that a long-pulse neutron source allows for the possibility of a wide range of flexible instrument concepts with variable resolution and dynamic range, tasks which invariably require the implementation of pulse-modulation techniques in the time domain, particularly for high-resolution applications. We also consider in some detail yet-to-be-tapped opportunities in the use of shorter proton pulses, characterised by a duration commensurate with typical moderation times at spallation sources.

1. Introduction

The *European Spallation Source* (ESS) project [22, 23] has undertaken the construction of the world's most intense neutron source for the study of condensed matter, including applications in physics, chemistry, geology, biology, materials discovery, and engineering [24]. The Spanish contribution to the ESS project is currently managed by the *ESS-Bilbao Consortium* (ESSB) and comprises both basic enginnering tasks as well as the delivery of machine components [4]. Within the latter, ESSB is actively engaged in the design of the future ESS instrument suite. In this context, the aim of this contribution is to present basic criteria associated with the novel use of long neutron pulses, with a view to guiding and facilitating subsequent efforts in instrument design and optimization. To this end, we examine the merits (and limitations) of millisecond (ms) neutron pulses produced by third-generation spallation neutron sources such as ESS. The considerations described herein constitute a natural extension of previous studies, some of which were aimed at the construction of a medium-flux source in Spain to produce neutron pulses up to durations of 1.5 ms, a pulse length compliant with the ESS 2009 technical specification (see, e.g., Refs. [5, 6, 7]). These results can be easily extrapolated to the current (and longer) temporal width of 2.86 ms envisaged for ESS, or to other accelerator-driven systems aimed at maximising the production of pulsed neutron beams, particularly in the sub-thermal (cold) regime [8].

2. Useful neutrons within a pulse

In accelerator-driven neutron sources, the pulse length reaching a given instrument depends on both the length of the particle pulse impinging the neutron-production target (typically protons) as well as on the time taken by thermalization processes at the moderators to slow down spallation (MeV) neutrons to usable energies, typically below 1 eV. In this spirit, a pulsed source may be classified as long- or short-pulse depending upon which temporal width is longer. Long(short)-pulse sources are those where the temporal duration of the accelerator pulse becomes significantly longer(shorter) than the time required for neutron moderation down to thermal energies [9].

By their very nature, neutron scattering techniques with present-day technologies remain flux-limited [24, 10]. Although neutron-production reactions can yield total fluxes comparable to those characteristic of continuous reactor sources, the resulting *useful* flux at thermal energies is an extremely inefficient process, a disadvantage that can be partially mitigated via recourse to multiplexing schemes, i.e., the synchronous use of a broad band of incident neutron wavelengths within a pulse [11, 12]. In this context, instrument design is primarily driven by the underlying scientific requirements. As on steady-state sources, the use of monochromatic incident beams at pulsed sources by means of velocity-selection devices is typically restricted to spectroscopic applications requiring wide surveys of momentum and energy transfers – so-called direct-geometry spectrometers. In this particular case, beam monochromatization before use translates into an effective re-definition of the source. Aside from the above, the most efficient exploitation of pulsed neutron beams is predicated upon the use of a wide range of incident wavelengths and time-of-flight (TOF) techniques to isolate the scattering response of a given neutron wavelength. Such is the case of pulsed neutron diffraction and so-called inverted-geometry spectrometers. In the latter case, energy selection is performed after use and, again, TOF methods are used to determine the corresponding energy and momentum transfers associated with a given incident neutron wavelength. As neutron scattering cross sections at thermal energies are typically independent of incident energy, both direct- and indirect-geometry spectrometers provide access to the same observables associated with the structure and dynamics of the system under investigation [13, 14]. Judicious instrument design in the latter case also allows for the simultaneous implementation of high-resolution neutron diffraction in backscattering geometry, as currently implemented in the low-energy IRIS & OSIRIS spectrometers [15, 16], and the thermal and epithermal spectrometers TOSCA [17, 18, 19] and VESUVIO [20], respectively, all located at the ISIS Pulsed Neutron and Muon Source, Rutherford Appleton Laboratory, United Kingdom.

In any of the cases presented above, the ultimate useful flux associated with a given spectral resolution and dynamic range at a pulsed neutron source depend on a number of factors which are examined below.

3. Temporal response, spectral resolution, and dynamic range

In addition to the total neutron flux at the point of use, an analysis of the temporal pulse line shapes for a given neutron wavelength is an interesting exercise in its own right in order to examine how these may be related to optimal neutronic performance. During the time lapse of the proton pulse, neutrons of different energies will emerge from the moderator roughly at the same time. As they travel away from the source, this initial pulse will spread out in time and energy by virtue of the inverse relationship between neutron velocity and wavelength. As a result, the pulse leading edge at a given distance from the source will contain a larger fraction of faster neutrons. Traditionally, the analysis of temporal line shapes has been restricted to the definition of the time-integrated neutron flux at a given wavelength and its associated temporal width, typically defined as a full-width-at-half-maximum (FWHM) in the time domain. Such FWHM tends to increase monotonically with wavelength, a dependence which conforms to physical intuition as the attainment of colder wavelengths requires on average a higher number of collisions with the moderating medium. For short-pulse neutron sources, wavelength-dependent FWHMs exhibit a (roughly) linear dependence on neutron wavelength in the thermal range, leading to a constant relative wavelength resolution [12]. This feature remains one of the primary virtues of both diffraction and indirect-geometry instrumentation at short-pulse sources where intrinsic FWHMs range from a few μsec at thermal wavelengths to 150-300 μsec for the coldest wavelengths approaching 10 Å. For a long-pulse source, additional pulse-modulation techniques are invariably needed to attain a comparable energy at a given distance from the source. In either case, a sharp and symmetric temporal line shape represents the best-case scenario in order to optimise useful neutron flux and spectral resolution at a given wavelength.

To examine deviations from the above scenario, we can extend previous treatments of neutron-pulse temporal line shapes in order to quantify deviations from a perfectly symmetric pulse around its most-probable (peak) value. In addition to the intrinsic FWHM, we can define Width and Line shape Asymmetry Parameters (WAP and LAP, respectively) as follows

$$WAP(\lambda) = \frac{HWHM_{right}(\lambda)}{HWHM_{left}(\lambda)} - 1 \tag{1}$$

and

$$LAP(\lambda) = \frac{\int_0^{\infty} \Phi(t,\lambda)dt}{2\int_0^{t_{max}} \Phi(t,\lambda)dt} - 1 \tag{2}$$

where t_{max} stands for the most-probable time of the time- and wavelength-dependent neutron flux $\Phi(t,\lambda)$ and HWHM denotes the half-width-at-half-maximum of $\Phi(t,\lambda)$ on either side of t_{max}. With these definitions, the WAP is most sensitive to a temporal asymmetry of $\Phi(t,\lambda)$ around t_{max} whereas the LAP provides an additional measure of long tails on either side of $\Phi(t,\lambda)$. Both WAP and LAP are positive(negative) depending on whether the time-dependent neutron flux is higher after(before) t_{max}. Positive WAP or LAP values are typical of short-pulse sources. Figures 1a and 1b show the FWHM, WAP, and LAP for two possible operational modes originally envisaged in the context of the ESSB project [6]. A few remarks on these results are in order. For the long pulse (1.5 ms), we observe a sharp transition for all three line shape parameters at around 1-2 Å. Following this step, all parameters remain roughly constant. The FWHM suffers a rather modest increase from 1.5 to 1.65 ms, whereas both WAP and LAP start at their minimal values at short wavelengths (undermoderated regime) and become markely more symmetric beyond 2 Å. These considerations are in line with the notion that neutron pulses in this regime are largely dictated by the temporal structure of the proton pulse. Conversely, they also tell us that suitable schemes could be further developed to modify the temporal response of neutron pulses by changing the proton-pulse profile. For the shorter pulse (0.1 ms), the situation is markedly more complex as a consequence of the more similar timescales of the proton pulse and intrinsic moderation times. Both FWHM and WAP display a similar wavelength dependence, indicative of longer moderation times for colder neutrons. Both LAP and WAP become positive at 1 and 3 Å, respectively, indicative of the emergence of broad tails in the time-dependent flux. Interestingly, the LAP levels off to a constant value at much shorter wavelengths than the WAP indicating that temporal-pulse degradation is already important for energies sensibly above typical moderator temperatures. We also note that this proton-pulse width (0.1 ms) is intermediate between the short-pulse limit of second-generation spallation sources like ISIS at the Rutherford Appleton Laboratory, the SNS at Oak Ridge National Laboratory, or the MLF at J-PARC (μsec) and the much-longer pulses at ESS (few ms). Given the similarities in timescales between proton-pulse widths and moderation times, it is worthwhile noting that this regime represents (at least in principle) a suitable starting point for the optimal delivery of *both* incident flux and spectral resolution. These considerations have been examined in more detail in

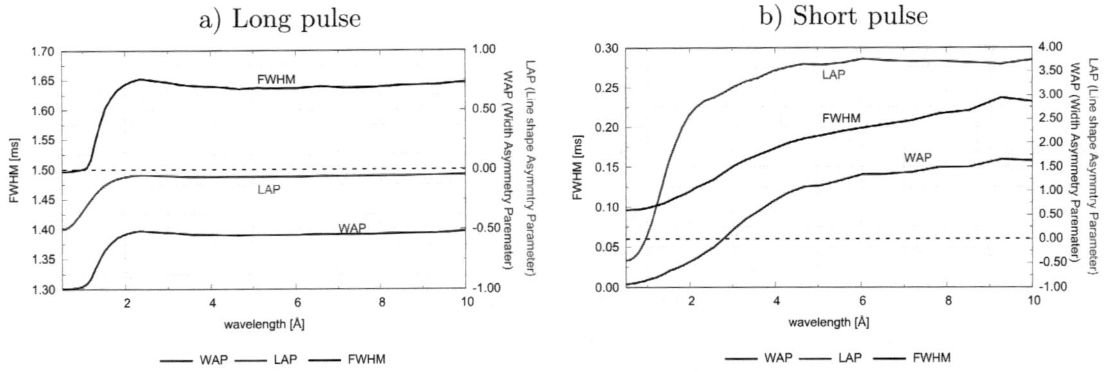

Figure 1. FWHM, WAP, and LAP for two possible operational modes originally envisaged in the context of the ESSB project: (a) long-pulse mode (1.5 ms proton pulse at a repetition frequency of 20 Hz); (b) short-pulse mode (0.1 ms proton pulse, 50 Hz). For further details, see the main text.

Refs. [23, 6], where it was concluded that a 0.1 ms/50 Hz proton pulse offers a wider spectrum of opportunities for neutron-scattering techniques than its long-pulse counterpart (1.5 ms/20 Hz).

On the basis of our discussion above, we can also use temporal pulse widths to define a characteristic wavelength-dependent resolution $R(\lambda)$ which can be written as [6]

$$R(\lambda) = \frac{3.96}{\lambda[\mathring{A}]} \cdot \frac{\Delta t[ms]}{L[m]} \tag{3}$$

where Δt is the neutron pulse width and L is the length of the instrument. Figure 2 shows the minimum instrument length at ESS required to reach a certain value of $R(\lambda)$. From these data, we infer that medium-to-high resolution would only be within reach via the use of flight paths of hundreds of meters. This condition makes the use of pulse-modulation and pulse-shaping techniques a must in order to achieve high resolution at a reasonable cost.

Unlike steady-state sources, pulsed neutron sources are unique in that it becomes possible to exploit a wide dynamic range within each neutron pulse using TOF techniques. This already-mentioned multiplexing advantage can lead to order-of-magnitude gains in count rate and also explains why the peak neutron flux is a more appropriate figure of merit to assess instrument performance, as opposed to analogous time-averaged quantities [11]. The dynamic range $\Delta\lambda$ is both dependent on the distance to the source (as the spectral resolution) as well as on the repetition rate of the source (f), as shown by the following equation [6]

$$\Delta\lambda[\mathring{A}] = \Big(\frac{10^3}{f[Hz]} - \Delta t[ms]\Big) \cdot \frac{3.96}{L[m]} \tag{4}$$

Figure 3 shows the requisite instrument lengths needed to attain a certain dynamic range for a given value of the spectral resolution. Also in this case, it reinforces the need to implement pulse-modulation and shaping techniques, otherwise the raw ESS pulse will never achieve high resolution while keeping a useful dynamic range. This requirement will necessarily reduce the number of useful neutrons per pulse to a fraction of the available total, although it also represents a timely opportunity for the further development of these technologies beyond the present state-of-the-art.

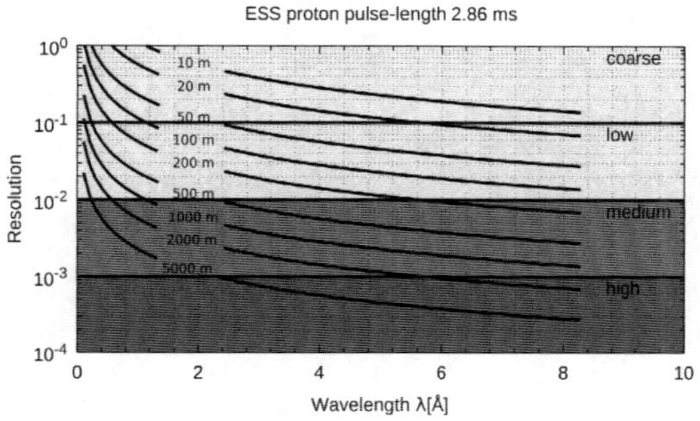

Figure 2. Spectral resolution for a number of instrument lengths at ESS. As a guide, qualitative resolution ranges are also shown.

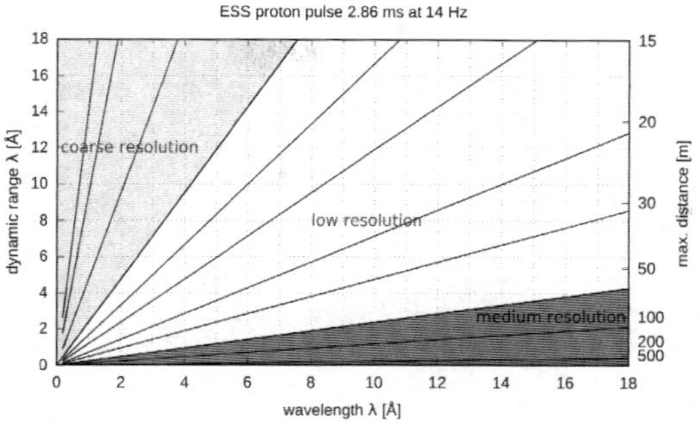

Figure 3. Available dynamic range and spectral resolution for a given instrument length at ESS. For further details, see the text.

4. Conclusions and outlook

In this work, we have presented a number of guiding criteria for instrument selection at state-of-the-art spallation neutron sources, where the proton-beam pulse can range from hundreds of μsec to a few ms. These considerations represent a natural continuation of our efforts to select optimal proton-pulse characteristics in the context of the ESSB project, yet at the same time are equally applicable to wider initiatives to develop third-generation neutron sources in the foreseeable future, ESS being an important and timely case in point. In addition to considering the maximisation and subsequent delivery of *sheer* incident flux, we have extended the traditional analysis of temporal neutron-pulse line shapes to quantify the overall width and underlying asymmetry close and away from the most-probable value. From this analysis, we conclude that medium-to-high-resolution applications with long (ms) pulses would require the extensive implementation of novel pulse-shaping and modulation techniques, leading to an unavoidable loss in useful flux. We also find that shorter proton pulses of the order of 0.1 ms are naturally matched to typical moderation times of sub-thermal neutrons at a spallation target and, therefore, they could offer a convenient starting point for the effective implementation of a wide range of neutron-scattering techniques at a reasonable cost.

Acknowledgments

We would like to express our gratitude to the *ISIS Molecular Sprectroscopy Group* for their hospitality and guidance during the course of this project. This work would not have been possible without the support of the computing infrastructure of the i2BASQUE Academic Network and the support of the ESS-DMSC Computing Centre.

References

[1] Url: *www.europeanspallationsource.se*

[2] Peggs S *et al.* (Eds), *ESS Technical Design Report* (European Spallation Source, Lund, 2013). Url: *eval.esss.lu.se/cgi-bin/public/DocDB/ShowDocument?docid=274*

[3] Fernandez-Alonso F and Price D L (Eds), *Neutron Scattering - Fundamentals* (Academic Press, New York, 2013).

[4] Url: *www.essbilbao.org*

[5] Bermejo f J and Sordo F (Eds), *Technical Design Report: ESS-Bilbao Target Station* (Consorcio ESS-Bilbao, Bilbao, 2013). Url: *www.essbilbao.org:8080/ESSBilbao/en/copy_of_LAN_TDR.pdf*

[6] de Vicente J P, Fernandez-Alonso F, Sordo F, and Bermejo F J, *Rutherford Appleton Laboratory*

Technical Report RAL-TR-2013-016 (Science & Technology Facilities Council, Didcot, 2013). Url: *epubs.stfc.ac.uk/work/11230408*

[7] Sordo F, Fernandez-Alonso F, Terrón S, Magán M, Ghiglino A, Martínez F, Bermejo F J, and Perlado J M, *Phys. Procedia* **60** 125 (2014).

[8] Vivanco R, Ghiglino A, de Vicente J P, Sordo F, Terrón S, Magán M, Perlado J M, and Bermejo F J, *Nucl. Instr. Meth. Phys. Res. A* **767** 176 (2014).

[9] Bermejo F J and Sordo F, *Neutron Sources*, in Fernandez-Alonso F and Price D L (Eds), *Neutron Scattering - Fundamentals* (Academic Press, New York, 2013), Chapter 2.

[10] Pynn R, *Neutron scattering - A Primer* (Los Alamos Science, Los Alamos, 1990). Url: *www.mrl.ucsb.edu/ seshadri/RogerPynn-Tutorial/Pynn-NeutronPrimer.pdf*

[11] Windsor C G, *Pulsed Neutron Scattering* (Taylor & Francis, London, 1981).

[12] Arai M, *Experimental Techniques*, in Fernandez-Alonso F and Price D L (Eds), *Neutron Scattering - Fundamentals* (Academic Press, New York, 2013), Chapter 3.

[13] Price D L and Fernandez-Alonso F, *An Introduction to Neutron Scattering*, in Fernandez-Alonso F and Price D L (Eds), *Neutron Scattering - Fundamentals* (Academic Press, New York, 2013), Chapter 1.

[14] Dawidowski J, Granada J R, Santisteban J R, Cantargi F, and Rodríguez Palomino L A, *Neutron Scattering Lengths and Cross Sections*, in Fernandez-Alonso F and Price D L (Eds), *Neutron Scattering - Fundamentals* (Academic Press, New York, 2013), Appendix.

[15] Demmel F, Garcia-Sakai V, Mukhopadhyay S, Parker S F, Fernandez-Alonso F, Armstrong J, Bresme F, de Vicente J P, Sordo F, Bermejo F J, Salzmann C G, and McLain S E, *Rutherford Appleton Laboratory Technical Report RAL-TR-2014-014* (Science & Technology Facilities Council, Didcot, 2014). Url: *epubs.stfc.ac.uk/work/12271760*

[16] Demmel F, McPhail D, Crawford J, Maxwell D, Pokhilchuk K, Garcia Sakai V, Mukhopadhyay S, Telling M T F, Bermejo F J, Skipper N T, and Fernandez-Alonso F, *Eur. Phys. J. Web of Conferences* **83** 03003 (2015).

[17] Rudic S, Ramirez-Cuesta A J, Parker S F, Fernandez-Alonso F, Pinna R S, Gorini G, Salzmann C G, McLain S E, and Skipper N T, *Rutherford Appleton Laboratory Technical Report RAL-TR-2013-015* (Science & Technology Facilities Council, Didcot, 2013). Url: *epubs.stfc.ac.uk/work/11216706*

[18] Parker S F, Fernandez-Alonso F, Ramirez-Cuesta A J, Tomkinson J, Rudic S, Pinna R S, Gorini G, and Fernández Castañón J, *J. Phys.: Conf. Ser.* **554** 012003 (2014).

[19] Pinna R S, Rudic S, Parker S F, Gorini G, and Fernandez-Alonso F, *Eur. Phys. J. Web of Conferences* **83** 03013 (2015).

[20] Seel A G, Krzystyniak M, and Fernandez-Alonso F, *J. Phys.: Conf. Ser.* **571** 012006 (2014).

[21] Krzystyniak M, Adams M A, Lovell A, Skipper N T, Bennington S M, Mayers J, and Fernandez-Alonso F, *Faraday Discuss.* **151** 171 (2011).

[22] Url: *www.europeanspallationsource.se*

[23] Peggs S *et al.* (Eds), *ESS Technical Design Report* (European Spallation Source, Lund, 2013). Url: *eval.esss.lu.se/cgi-bin/public/DocDB/ShowDocument?docid=274*

[24] Fernandez-Alonso F and Price D L (Eds), *Neutron Scattering - Fundamentals* (Academic Press, New York, 2013).

Journal of Physics: Conference Series **663** (2015) 012012 doi:10.1088/1742-6596/663/1/012012

Low Temperature Magnetic Ordering of the Magnetic Ionic Plastic Crystal, Choline[$FeCl_4$]

I. de Pedro*[1], A. García-Saiz[1], D. Andreica[2], L. Fernández Barquín[1], M.T. Fernández-Díaz[3] J. A. Blanco[4] A. Amato[5] and J. Rodríguez Fernández[1]

[1] CITIMAC, Facultad de Ciencias, Universidad de Cantabria, 39005 Santander, Spain.

[2] Faculty of Physics, Babes-Bolyai University, 400084 Cluj-Napoca, Romania

[3] Institut Laue-Langevin, BP 156X, F-38042 Grenoble Cedex, France

[4] Departamento de Física, Universidad de Oviedo, 33007 Oviedo, Spain

[5] Laboratory for Muon Spin Spectroscopy, Paul Scherrer Institut, 5232 Villigen PSI, Switzerland

E-mail: depedrovm@unican.es

Abstract. We report on the nature of the low temperature magnetic ordering of a magnetic ionic plastic crystal, Choline[$FeCl_4$]. This investigation was carried out using heat capacity measurements, neutron diffraction experiments and muon spin relaxation (μSR) spectroscopy. The calorimetric measurements show the onset of an unusual magnetic ordering below 4 K with a possible second magnetic phase transition below 2 K. Low temperature neutron diffraction data reveal a three dimensional antiferromagnetic ordering at 2 K compatible with the previous magnetometry results. The analysis of μSR spectra indicates a magnetic phase transition below 2.2 K. At 1.6 K, the analysis of the shape of the μSR spectra suggests the existence of an additional magnetic phase with features of a possible incommensurate magnetic structure.

1. Introduction

Plastic crystal materials were described by Timmermans in the 1960s. These compounds, composed entirely of ions, display an intermediate phase between the solid and the liquid state called plastic phase. It is reached by one or more solid–solid phase transitions on warming the fully ordered crystalline phase from low temperatures [1]. This phase is characterized by a state presenting both long-range order and short-range disorder. Such as short-range is originated by rotational motions of the molecules. The plastic behaviour, with a negligible vapour pressure and high ionic conductivity at room temperature, reinforce their interest as solid state ion conductors [2, 3].

The introduction of structural functionalities on the cationic or anionic part has eased the design of new plastic crystal phases with targeted properties. Transition metal-containing plastic phase, such as organic inorganic plastic crystals (OIPCs) [4] could be auspicious new emergent materials, which will favourably combine the properties of ionic plastic crystals with the magnetism, originated from the metal incorporated in the complex anion. This is the case of the N,N,N-trimethylethanolammonium tetrachloroferrate (III), Choline[$FeCl_4$], which combines a three-dimensional magnetic ordering with a high solid-state-conduction [5], becoming a promising candidate for its use as electrolyte in solid state lithium batteries.

A structural and magnetic investigation of Choline[$FeCl_4$] was carried out using Raman spectroscopy and time-resolved X-ray thermodiffractometry. The most striking features are an enhancement of the rotational motion of $[Choline]^+$ and $[FeCl_4]^-$ anions with increasing temperature up to the melting point, 490 K, and the presence of a three-dimensional magnetic ordering below 4 K [5]. In this contribution we focus on the nature of the low temperature magnetic ordering of this compound combining heat capacity measurements, neutron diffraction data and muon spin relaxation (μSR) spectroscopy.

2. Results and discussion

Heat capacity (Cp) measurements were carried out with a standard QD PPMS device by a relaxation method using a two-tau model. Data were collected from 2 to 10 K at several magnetic fields between 0 and 50 kOe (Figure 1). In absence of an external magnetic field, when the temperature is lowered, the Cp value starts to increase below 5 K, showing a shoulder around 3.4 K. Then, a strong increase down to 2 K (ΔC_p=13 J/Kmol) is detected. The first anomaly (3.4 K) does not have the typical behaviour of a second-order transition, but it can be associated with the establishment of a three-dimensional antiferromagnetic ordering, in good agreement with magnetic susceptibility measurements [5]. In addition, the strong increase down to 2 K could be related to a possible additional magnetic transition at lower temperatures. By applying magnetic fields, the shoulder remains up to 30 kOe, but vanishes for larger fields. This vanishing shoulder corroborates the presence of the weak magnetocrystalline anisotropy detected in the magnetization data [5]. Furthermore, the second magnetic anomaly decreases its magnitude as the magnetic field increases, and probably shifting to lower temperatures, as is expected for an antiferromagnetic ordering. Therefore, neutron and muon studies were performed in order to understand the origin of this unusual magnetic behaviour.

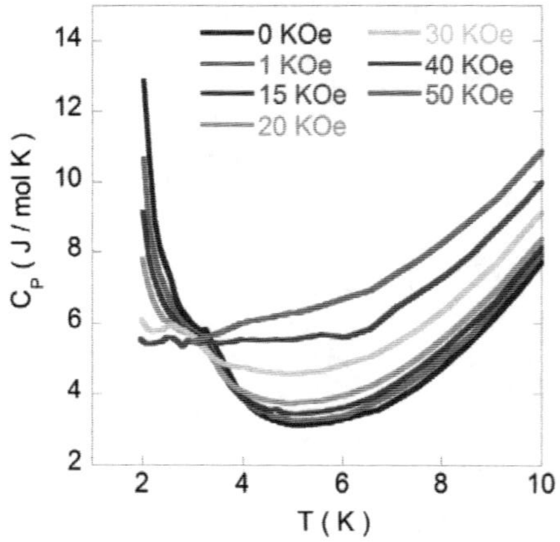

Figure 1. Temperature dependence of the heat capacity of Choline[$FeCl_4$] under external magnetic fields ($0 \leq H \leq 50$ kOe).

Powder neutron diffraction measurements were performed on the D1B diffractometer, at the Institute Laue-Langevin of Grenoble, using a neutron wavelength of 2.525 Å. About 3 g of Choline[$FeCl_4$] was employed in the neutron diffraction experiments, placed in a cylindrical vanadium container and held in a liquid helium cryostat. The temperature evolution of the D1B patterns between 2 and 10 K shows the presence of extra elastic intensity at temperatures below around 4 K, corroborating the onset of a magnetic ordering. A comparative view of the patterns above (10 K) and below (2 K) the magnetic ordering temperature is shown in Figure 2 (a). The most significant feature being the presence of an extra magnetic signal ($2\theta \approx 15°$), which could be attributed to the establishment of a three-dimensional antiferromagnetic ordering. A satisfactory monoclinic solution with unit cell parameters of a = 15.376(1) Å, b = 15.234(1) Å, c = 8.742(1) Å, β = 99.20(1) and V = 2021.52 (1) Å^3 was found using the WINPLOTR programme. The additional magnetic peak at 2 K was indexed with the propagation vector k = (0, 0.17, 0) indicating that the magnetic structure is incommensurate. The experimental, calculated and difference diffraction profiles are displayed in Figures 2 (b). However, with only one visible magnetic peak, we cannot determine univocally the magnetic propagation vector. As observed in other hybrid organic–inorganic materials based in tetrahaloferrate cation such as Dimim[FeX_4] [6, 7, 8], Edimim[FeX_4] [9] or Emim[FeX_4] (X = Cl and Br) [10, 11] the magnetic structure should consist of an arrangement of iron metal complex groups, stabilizing the overall antiferromagnetic ordering without direct Fe··Fe interactions, but through the superexchange interactions, Fe-Cl··Cl-Fe.

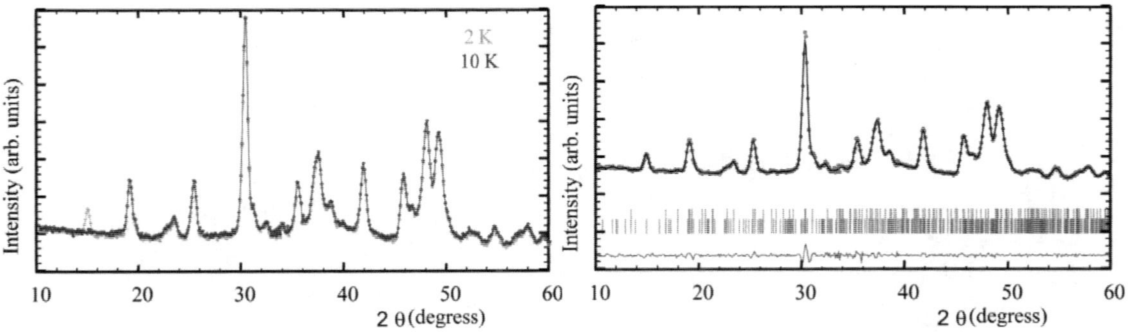

Figure 2. (a) Neutron patterns at 10 and 2 K for Choline[$FeCl_4$] in D1B difractometer. The (*) dot show the magnetic contributions. (b) Neutron diffraction profiles for Choline[$FeCl_4$] phase at 2 K. Positions of the magnetic (second row) profile are also presented.

The μSR experiments were performed in zero applied field at the Laboratory for Muon Spin Spectroscopy of the Paul Scherrer Institute, Switzerland using the GPS spectrometer in the temperature range 2–200 K. Above 4 K, the whole sample volume is in the paramagnetic phase. For the entire temperature range, the μSR data were fitted using the sum of an exponentially damped – Kubo-Toyabe depolarization function [12], first term in eq. (1), with a term describing the μ^+ behaviour in the magnetic phase:

$$A_{para} e^{-\lambda t}\left(\frac{1}{3}+\frac{2}{3}\left(1-\sigma^2 t^2\right)e^{-\frac{\sigma^2 t^2}{2}}\right)+A_{magn}P_{magn}(t) \quad (1)$$

σ can be expressed in terms of the interaction between the muon spin and the spins of the compound and the characteristics of the latter spins; $A_{magn}+A_{para}=A$ where A is a parameter called asymmetry that can be obtained from a calibration run, in the paramagnetic phase, in weak transverse field. The first term in Eq. (1) describes a double channel depolarization of the muon ensemble, in the paramagnetic phase: one arising from the fluctuating electronic magnetic moments, the exponential relaxation, and the other one from the static (for the time window of the μSR experiments) depolarization due to the nuclear magnetic moments, the so-called Kubo-Toyabe depolarization function. The second term of the sum in Eq. (1) describes the magnetic volume and has different T dependencies, which are dependent of the temperature range. $A_{magn}=0$ for $T>4\,\mathrm{K}$ while $A_{para}=0$ at lower temperatures.

As one can see from Figure 3, below 4 K, a fast relaxing component sets in, confirming the onset of a magnetic ordering detected by heat capacity and neutron powder diffraction measurements. Down to about 2.4 K, the "magnetic part" of the μSR spectra could be described by a purely exponential $A_{magn}P_{magn}(t)$, i.e. no oscillations are observed in the μSR spectra. The temperature dependence of A_{para} in this temperature range is displayed in the inset of Figure 3 (right). Below 2.4 K there is clear indication of a (broad) non-zero average field at the muon site: oscillations are observed in the μSR spectra. $A_{magn}P_{magn}(t)$ that fits best the μSR spectra has the following expression:

$$A_{magn}P_{magn}(t) = A_1 e^{-\lambda_1 t} J_0(2\pi\nu_1) + A_2 e^{-\lambda_2 t} J_0(2\pi\nu_2) + A_3 e^{-\lambda_3 t} \quad (2)$$

where $A_1 + A_2 + A_3 = A_{magn}$, $A_3 = \frac{A_1 + A_2}{2}$ and $J_0(2\pi\nu)$ is a 0'st order Bessel function.

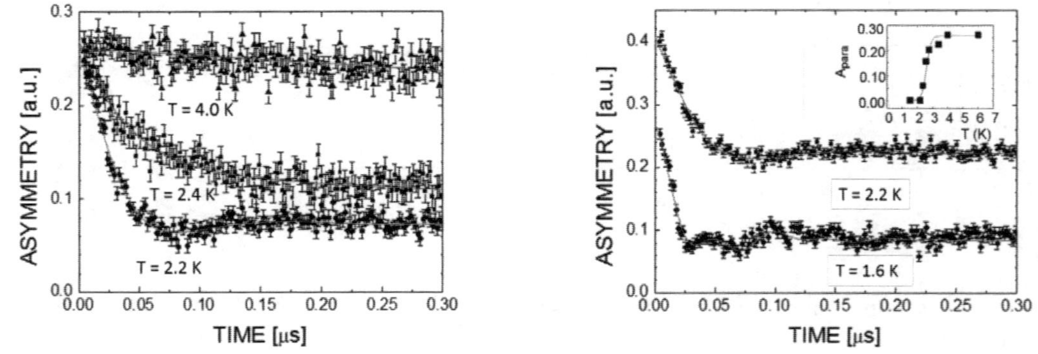

Figure 3. μSR spectra recorded on Choline[$FeCl_4$] at different temperatures between 1.6 and 4 K. In the left figure the 2.2 K spectra was shifted up for clarity. The solid lines represent the fitting according the Eq. (1) and (2). The inset shows the temperature dependence of A_{para}. The solid line is a guide to the eye.

Usually, a Bessel function fit of the μSR spectra is an indication of an incommensurate magnetic ordering (spin density wave). However, another fully plausible explanation would be the existence of a wide, non-symmetric field distribution at the muon site. Therefore, μSR results alone cannot decide whether the magnetic ordering of Choline[$FeCl_4$] below 2.2 K is incommensurate or not.

3. Conclusions

Heat capacity measurements show that the magnetic ionic plastic crystal Choline[$FeCl_4$] exhibits two successive magnetic transitions below 4 K. Neutron diffraction data confirms the antiferromagnetic nature of the long-range magnetic order whereas muon spin relaxation spectroscopy experiments, although not conclusive, are compatible with a possible additional incommensurate magnetic structure below 2 K.

4. Acknowledgements

Part of this work was carried out at the Swiss Muon Source (PSI, Villigen, Switzerland) and Institute Laue-Langevin (ILL) of Grenoble and was partially financially supported by MEC research projects MAT2011-27573-C04 and Romanian UEFISCDI Project No. PN-II-ID-PCE-2011-3-0583 (85/2011)

5. References

[1] Timmermans J 1961 *J. Phys. Chem. Solids* **18** 1
[2] Howlett P. C, Shekibi Y, MacFarlane D. R and Forsyth M 2009 *Adv. Eng. Mater.* **11** 1044
[3] MacFarlane D. R, Huang J and Forsyth M 1999 *Nature* **402** 792
[4] Lee M, Choi U. H, Wi S, Slebodnick C, Colby R. H and Gibson H. W 2011 *J. Mater. Chem.* **21** 12280
[5] de Pedro I, García-Saiz A, González J, de Larramendi I. R, Rojo T, Afonso C. A and Rodríguez Fernández J. 2013 *Phys. Chem. Chem. Phys.* **15(30)** 12724
[6] García-Saiz A, de Pedro I, Migowski P, Vallcorba O, Junquera J, Blanco J. A and Rodríguez Fernández, J 2014 Inorg. Chem. **53(16)** 8384
[7] García-Saiz A, Migowski P, Vallcorba O, Junquera J, Blanco J. A, González J. A, Fernández-Díaz M. T, Rius J, Dupont J, Rodríguez Fernández J and de Pedro I 2014 *Chem. Eur. J.* **20** 72
[8] García-Saiz A, de Pedro I, Blanco J A González J. A and Rodríguez Fernández, J 2013 *J. Phys. Chem. B* **117** 3198
[9] García Saiz A, de Pedro I, Vallcorba O, Migowski P, Campo I, Barquin, L F, Abrahams I, Motevalli M, Dupont J, Gonzalez J A, Rodríguez Fernández, J 2015 *RSC Adv.* **5** 60835
[10] de Pedro I, Rojas D. P, Albo J, Luis P, Irabien A, Blanco J A and Rodríguez Fernández, J 2010 *J. Phys.: Condens. Matter* **22(29)** 296006
[11] de Pedro, I, Rojas D. P, Blanco J A and Rodríguez Fernández, J 2011 *J. Magn. Magn. Mater.* **323** 1254
[12] Hayano R. S, Uemura Y. J, Imazato J, Nishida N, Yamazaki T and Kubo R 1979 *Phys. Rev. B* **20(3)** 85

Mobility of dislocations and grain boundaries controlled by the order degree in iron-based alloys

D Gargicevich[1], O A Lambri[1*], J I Pérez-Landazábal[2,3], V Recarte[2,3], F G Bonifacich[1], G J Cuello[4], V Sánchez Alarcos[2]
[1]Laboratorio de Materiales, Escuela de Ingeniería Eléctrica, Centro de Tecnología e Investigación Eléctrica, Facultad de Ciencias Exactas, Ingeniería y Agrimensura, Universidad Nacional de Rosario - CONICET, Av. Pellegrini 250, 2000, Rosario, Argentina.
[2]Departamento de Física, Institute for Advanced Materials (INAMAT) Universidad Pública de Navarra, Campus de Arrosadía s/n, 31006 Pamplona, Spain.
[3]Institute for Advanced Materials (INAMAT), Universidad Pública de Navarra, Campus de Arrosadía, 31006 Pamplona, Spain
[4]Institute Laue-Langevin, 71 Av. des Martyrs, B.P. 156, 38042 Grenoble Cedex 9, France.

*E-mail: olambri@fceia.unr.adu.ar

Abstract Neutron diffraction and mechanical spectroscopy studies were performed in Fe-Al-Cr, Fe-Al-Si and Fe-Si alloys. Results show that in ordered alloys independently of the type of order, $D0_3$ or B2, the mobility of dislocations and grain boundaries is markedly reduced. It is revealed through small values both of damping and of the strength of the amplitude dependent damping (S). Moreover, the behaviour of S as a function of temperature is a suitable tool for determining the order changes as a function of temperature.

1. Introduction

In ordered lattices, dislocations should move in pairs, giving rise to the so called superdislocations, which leads to a decrease in their mobility regarding to the disordered lattice. Then, the recovery of quenched-in-defects and -dislocations has a slow rate; affecting also to the grain boundary mobility. Neutron diffraction (ND) and mechanical spectroscopy (MS) studies were coupled in this work for correlating the order degree and its effect on the mobility of dislocations and grain boundaries in Fe-Al-Cr, Fe-Al-Si and Fe-Si alloys. Indeed, MS is a very sensitive technique for studying the interaction processes between defects in materials [1].

2. Experimental

Samples of two iron based ternary systems, Fe-Al-Cr and Fe-Al-Si, and binary Fe-Si have been studied in this work. Sample composition, in at.% , are: Fe-25Al-8Cr, Fe-25Al-25Cr, Fe-12Al-12Si and Fe-10Si; respectively labelled as: TCr8, TCr25, TSi12 and B10. Sample details are described in Ref. [2-4] for ternary systems and in Ref. [5] for Fe-Si binary samples.

ND studies on ternary alloys were performed at the D1B powder diffractometer in the Institute Laue Langevin (ILL), Grenoble, France. The employed neutron wavelengths for Fe-Al-Cr and Fe-Al-Si alloys were λ = 2.52Å and 1.28Å, respectively. For Fe-Si alloys ND studies were performed at the D20 installation at the ILL, using a neutron wavelength of λ = 1.3Å. In situ measurements, in both diffractometers, were performed during heating at 3K/min from room temperature up to different maximum temperatures under vacuum (10^{-2} Pa).

MS, measurements were performed in an inverted torsion pendulum at frequencies close to 1 Hz under vacuum (10^{-5}Pa). Damping and square frequency, were measured with an error less than 2%. The measurements were carried out during subsequent heating and cooling runs, both at 1K/min, on the same specimen. Amplitude dependent damping (ADD), i.e. damping as a function of the maximum strain on the sample, ε_0, was calculated by evaluating the derivative of the curve corresponding to the decaying areas of the torsional vibrations as a function of the period number by means of Chi-square fitting. The strength of ADD effects is measured through S parameter (S = $\Delta Q^{-1} / \Delta\varepsilon_0$).

3. Results and Discussions

Figures 1.a and 1.b show a three-dimensional picture (intensity, 2θ and temperature) of the phase evolution "in situ" observed at the diffractometer for samples TCr8 and TCr25, respectively. On heating, the TCr8 sample exhibits two consecutive order→disorder transitions: the first one a $D0_3 \leftrightarrow$ B2 transition at 833 K from a cubic structure (space group $Fm\bar{3}m$) ordered at the next-nearest neighbours to a cubic one ordered at nearest neighbours (space group $Pm\bar{3}m$) and a second disordering process at 973 K to a $Im\bar{3}m$ cubic structure, $D0_3 \leftrightarrow$ B2 transition. On the other hand, the TCr25 sample only exhibits the B2 ↔ bcc transition at 973 K.

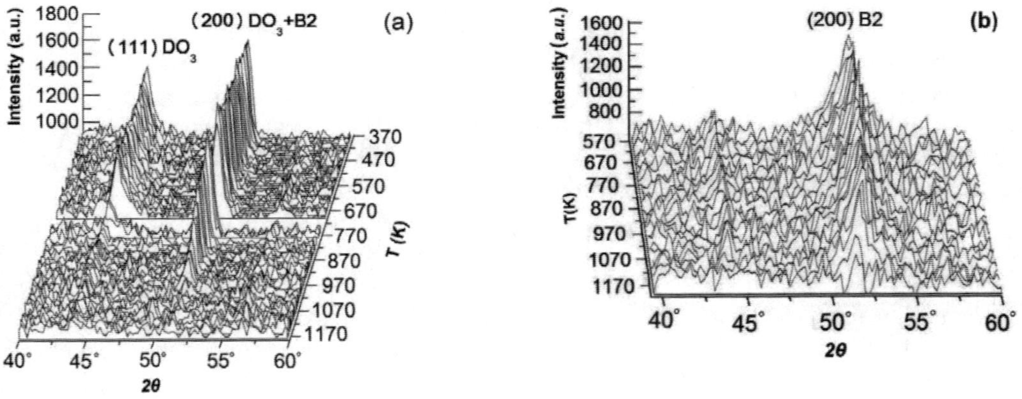

Figure 1: "In situ" neutron thermo-diffraction patterns measured during heating for as-quenched samples, (a) (TCr8) Fe-25Al-8Cr and (b) (TCr25) Fe-25Al-25Cr.

ND patterns at room temperature after the first "in situ" heating show differences in the degree of order between this annealing state and the as quenched one. For instance Fig. 2 shows the room temperature spectra for binary B10 sample, both in the as-quenched state and after annealing up to 1170 K. A small increase in the integrated area after thermal treatment above 1050 K can be observed indicating that the order in the as-quenched sample is not complete. No changes in the full width at half maximum of the diffraction peaks were found for all the group of studied samples, which indicate that changes in the grain size do not occur during heating; in agreement with the metallurgical observations [2, 3].

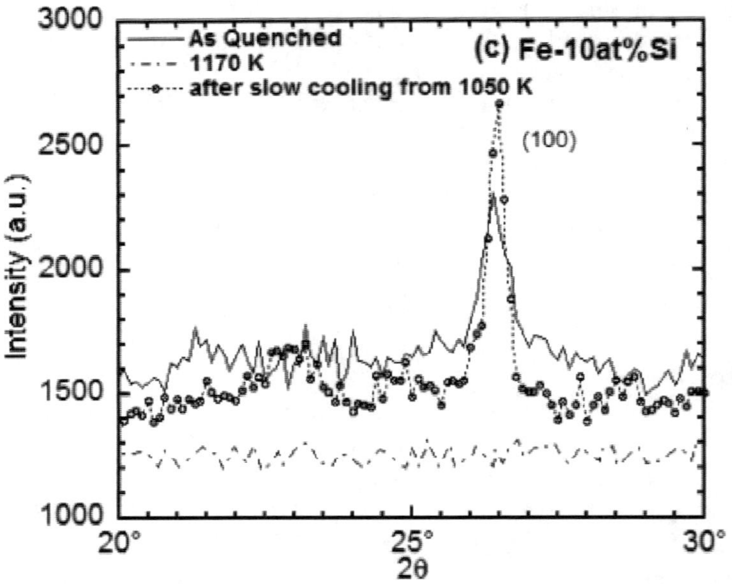

Figure 2. Neutron diffraction patterns for binary B10 sample at room temperature, as quenched state (continuous line), after the first heating (dots) and at 1170 K (dashed line)

The damping measured at increasing amplitudes is at first a constant, then increases sharply above amplitudes of around 10^{-6} [1, 6]. This behaviour has been explained by Granato and Lücke as due to the breaking away of dislocations loops from pinning impurities [7]. Under increasing applied stress, the lengths of dislocations between pinning points bow out until the dislocation escapes from pinning points. This mechanism of break-away of dislocation from weak pinning points give rise to the appearance of amplitude dependent damping (ADD). Nevertheless, ADD can be also used for studying heavy plastically worked materials by considering the average free length of dislocations [8]. Therefore, the order changes occurring in alloys as a function of temperature should be described by means of ADD since the decrease in the dislocation mobility in the ordered lattice. It leads to smaller bow outs and then smaller amount of breaking away processes due to the smaller swept areas by the dislocation movement. Indeed Figure 3 shows the behaviour of the S parameter as a function of temperature for the TCr8 sample, during the first heating from 600 K up to 1050 K. In addition, the evolution of the squared root of the intensities for the (200) and (111) reflections, which are proportional to the order parameter (η) for the $D0_3$ and B2 structures respectively, is also shown in Figure 3. S exhibits a slope change at the temperatures where the order state is modified. The first increase in S appears for the $D0_3 \rightarrow$ B2 transformation at 833K and the second one appears for the B2 $\rightarrow$ bcc transition at 973 K. In fact, dislocations should move in pairs in the ordered lattice, which leads to a decrease in their mobility. A decrease in the order degree enhances the mobility of the dislocation line, allowing sweep larger areas by bowing. It leads to a larger amount of weak pinning points inside the swept area, where the dislocation could escape during its movement, giving rise to the increase in S values. The sharp S increase above 973 K is promoted by the increase in the dislocation mobility in the disordered phase, which is in agreement with the transition B2 $\rightarrow$ bcc. Besides, the peak-shaped in S at around 890 K can be related to the change in mobility for dislocations between the transition $D0_3 \rightarrow$ B2.

Similarly to the behavior exhibited by sample TCr8, sample B10 shows a marked increase in S at the temperature where the (200) peak intensity begin dropping and the B2 $\rightarrow$ bcc transition develops (973 K), due to the increase in the dislocation mobility in the disordered phase, Figure 4. In addition, TSi12 sample show an increase in S for temperatures where the $D0_3$ order start to decrease, i.e. over around 860 K, Figure 4. Then, regardless of the sequence of order-disorder transitions observed the reduction of long-range order degree implies an increase of the damping linked to a higher mobility of dislocations.

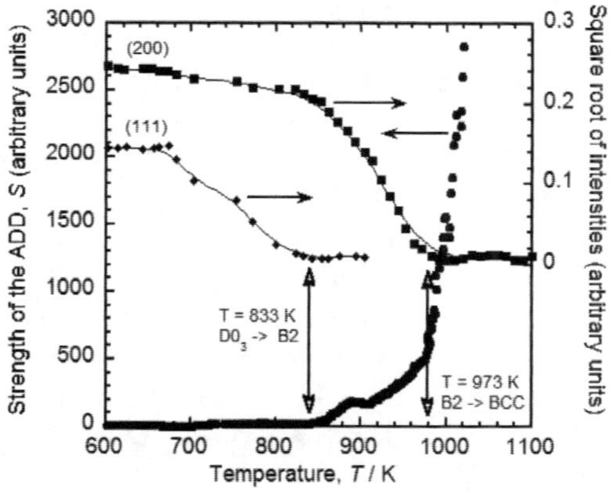

Figure 3: S as a function of temperature and square roots of intensities of the diffraction peaks for a type TCr8 sample. Vertical arrows indicate the transitions temperatures.

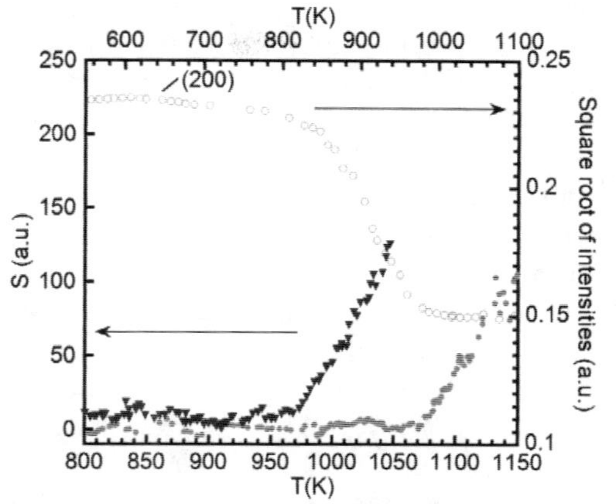

Figure 4: S (lower and left axis) and square root of intensities of the diffraction peaks (upper and right axis) for: TSi12 (circles) and B10 (triangles).

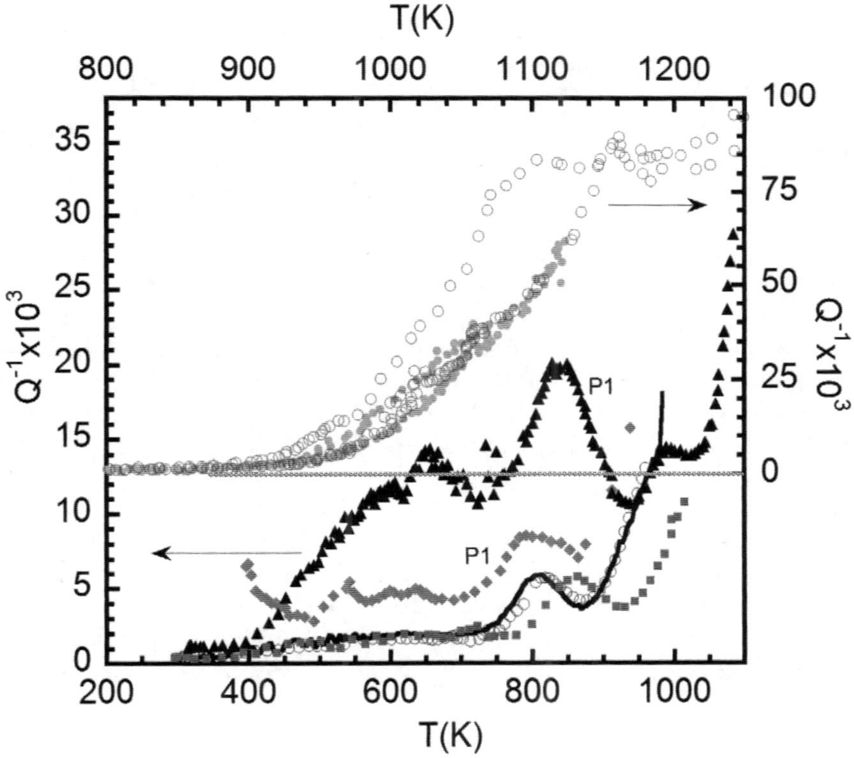

Figure 5: Lower and left axis: Damping spectra for a B10 sample. Sample in the as-quenched state: full rhombuses. After a heating run up to 973K: empty circles. After two heating runs up to 973K: full squares. After a heating run up to 1050K: full line. After a heating run up to 1273K: full triangles. Upper and right axis: Damping spectra for TSi12 (circles). Full circles: thermal cycle up to 1130K. Empty circles: thermal cycle up to 1200 K.

Journal of Physics: Conference Series **663** (2015) 012013 doi:10.1088/1742-6596/663/1/012013

In the other hand, Figure 5 shows the damping spectra for a B10 sample after successive thermal cycles performed "in situ" at the spectrometer. All spectra show a characteristic grain boundary relaxation peak, P1, around 800 K [5, 9]. The peak temperature of the maximum related to P1 moves towards higher temperatures during the cycles up to 973 K. In the second heating the shift is very small, being larger during the third run up in temperature. In fact, during the slow cooling from the different maximum temperatures, the B2 order restores and the final order degree is higher than the corresponding to the as-quenched alloy (see Figure 2), leading to an increase in the internal stresses [5, 9]. In contrast, when the sample was previously measured up to 1050K, the peak temperature of the maximum during the heating run is shifted towards smaller temperatures (full line). Indeed, during the heating up to 1050K the sample transforms according to the phase diagram. The B2 phase transforms to a bcc at 973K and for higher temperatures the sample is disordered and the superdislocations disappear. Then, the structure can relax internal stresses which were retained in the ordered B2 superlattice. Nevertheless, the damping spectrum changes strongly when the sample was thermally treated up to 1273 K (full triangles, Figure 3), due to a large recovery of defects in the disordered lattice. The damping behaviour after successive heating ramps depends strongly on the temperature limit achieved. Heating below the order-disorder temperature (973 K) restores the order degree increasing internal stresses. In the case of passing the disordering temperature the structure of the dislocations change drastically enhancing the relaxation of stresses.

The damping spectra measured for the TSi12 sample, during thermal cycles up to two different final temperatures are also shown in Figure 5 by shifting the base-line. The damping spectra increase monotonously with the temperature increase and no relaxation peaks have been found during thermal cycles up to 1130 K. The heating and cooling runs are similar with a small thermal hysteresis. However, by increasing the temperature above 1200 K during the heating process, it leads to the appearance of a damping peak during the subsequent cooling at around 1100 K [2, 9]. In addition, for TCr8 and TCr25, the damping behavior as a function of the thermal cycles resulted similar to the exhibited by sample TSi12, but the hysteretic behaviour develops after thermal cycles up to around 1050 K [3]. In fact, the appearance of the hysteretic behaviour on damping is controlled by the decrease in the order degree, which leads to an increase in the mobility of dislocations and grain boundaries, giving rise to the appearance of the grain boundary peak related to the solute during the cooling run [2, 3, 9].

Therefore, as it can be inferred from the above results, the behaviour of S as a function of temperature follows the changes in the order state during the warming of the sample, i.e. the modification of the order state is viewed at the temperature where it occurs. In contrast, usual damping measurements as a function of temperature (Figure 5) could reveal the appearance of order changes after successive heating and cooling runs up to temperatures over the order-disorder temperature transition.

4. Conclusions

In ordered alloys independently of the type of order, $D0_3$ or B2, the mobility of dislocations and grain boundaries is markedly reduced, which is shown through small values of both damping and S. An excellent correlation between the order behaviour and S was found. A decrease in the order degree leads to an increase in S, due to the dislocation mobility increases. In addition, when the order decreases or disappears after annealing, the grain boundary mobility also increases.

Acknowledgements

The authors thank the Institute Laue Langeving (ILL) - D1B and D20 facilities, for the allocated neutron beam time (Exp. 1-1-94 and CRG-1121), Prof. I. S. Golovin for the ternary samples and to NKK for providing the Fe-Si alloys. This work was partially supported by the CONICET, PIP 2098 and 0179, the PID-UNR; ING 290 and 453 and the Cooperation Agreement between the UPNA and the UNR.

Journal of Physics: Conference Series **663** (2015) 012013 doi:10.1088/1742-6596/663/1/012013

References

[1] Nowick A S and Berry B S, Anelastic Relaxation in Crystalline Solids, Academic Press, New York, 1972

[2] Lambri O A, Pérez-Landazábal J I, Cuello G J, Cano J A, Recarte V, Siemers C and Golovin I S 2009 *J. Alloys and Compounds* **468** 96.

[3] Lambri O A, Pérez-Landazábal J I, Recarte V, Cuello G J and Golovin I S 2012 *J. Alloys and Compounds* **537** 117.

[4] Golovin I S, Strahl A and Neuhäuser H 2006 *Int. J. Mater. Res.* **97** 1078.

[5] Lambri O A, Pérez-Landazábal J I, Cano J A and Recarte V 2004 *Mater. Sci. Eng. A* **370** 459.

[6] Friedel J, Dislocations, Addison-Wesley, Reading, 1967

[7] Granato A and Lücke K 1956 *J. Appl. Phys.* **27** 583.

[8] Bedford A J, Fuller P G and Miller D R 1972 *J. Nucl. Mater.* **43** 164.

[9] Lambri O A, Pérez-Landazábal J I, Cuello G J, Gargicevich D, Recarte V, Bonifacich F G, Giordano E D and Sánchez-Alarcos V 2014 *Neutron News* **25** 37.

Neutron and synchrotron studies of structure and magnetism of Shape Memory Alloys

J. M. Barandiarán[1]*, V. A. Chernenko[1,2], P. Lázpita[1], J. Gutiérrez[1], M. L. Fdez-Gubieda[1] and A. Kimura[3]

[1] BCMaterials and University of the Basque Country (UPV/EHU), Leioa , Spain
[2] Ikerbasque, The Basque foundation for Science, Bilbao, Spain
[3] Graduate School of Science, Hiroshima University, Hiroshima, Japan

*manub@we.lc.ehu.es

Abstract. Magnetic Shape Memory Alloys have attracted a large interest from the basic aspects of their structural and magnetic transitions, as well as from their possible technological applications as very rapid, contactless, high stroke actuators. The typical alloy composition for such materials is Ni-Mn-X (X= Ga, In, Sb, Sn, etc.). The electronic and magnetic structure of such alloys are on the base of the displayed properties and are object of intense study by all kind of advanced experimental and theoretical methods. Here we present a summary of recent diffraction experiments with neutron, X ray Magnetic Circular Dichroism (XMCD) and X ray Photoemission Spectroscopy (XPS) that are complementary to elucidate the electronic and magnetic character of these alloys and the changes occurring at the martensitic transformation. While rigid band models can account for most electronic features, magnetism is better described by a localized moment mainly concentrated at the Mn atoms.

1. Introduction: Magnetic Shape Memory Alloys, neutrons and synchrotron

Magnetic Shape Memory Alloys (MSMA), with typical compositions Ni-Mn-X (X= Ga, In, Sb, Sn, etc.) have raised large interest during the last years, due to their multifunctional properties. Such properties are based on the ferromagnetic thermoelastic Martensitic Transformation (MT) that produces a multivariant martensitic phase. Martensites have complicated crystallographic and magnetic structures [1-3]. The martensitic and accompanying transitions are also complex; it involves phonon softening modes mechanisms and can end in several modulated and non-modulated structures, very sensitive to the exact composition of the alloys. Such structures are sometimes mixed up in a given alloy. Different degrees of order among the different atomic species can also be present in the same sample and evolve as a function of time and annealing. In addition, different magnetic atoms are involved in the magnetism. For instance, the prototype stoichiometric Heusler compound Ni_2MnGa, and off stoichiometric NiMnGa alloys, the most studied ones, have localized 3d magnetic moments linked to Mn and Ni atoms, with possible ferro and antiferromagnetic contributions to the coupling [4, 5]. Probably some itinerant contribution related to s electrons, shared by all components, is present too.

MSMA have been studied by X rays, TEM, synchrotron radiation and neutron techniques [2, 3, 6-9]. Neutron diffraction is a well-established technique for crystal and magnetic structure studies. As compared with X-ray and electron diffraction, it presents a number of advantages, like: a) the nuclear interaction is described by point like potentials [10, 11]. Scattering from neutrons gives a special sensitivity to large diffraction angles because do not depend on the scattering angle. b) Neutron scattering factors are not proportional to the atomic number, as in x-rays and electrons, and can even be negative, so that there can be a large contrast between neighbor elements (figure 1). c) Neutron penetration in matter is very large as compared with x-rays and electrons and bulk samples can be fully explored by neutrons, in contrast with near surface analysis by the other techniques. Also especial equipment, as cryostats and furnaces or superconducting magnets, can be employed with little perturbation of the beam and diffraction pattern. d) Neutrons are sensitive to the magnetic moment of the atoms and are invaluable to study magnetic structures. e) Neutron mass is similar to the atomic one so the inelastic interaction with matter, Inelastic Neutron Scattering (INS), can excite phonons and magnons which are difficult to observe with other types of radiation.

Neutrons also have some issues as: a) Neutrons studies need a large amount of sample. At least one gram is required for efficient powder diffraction data acquisition. b) There is no direct interaction with the electronic structure of the solid.

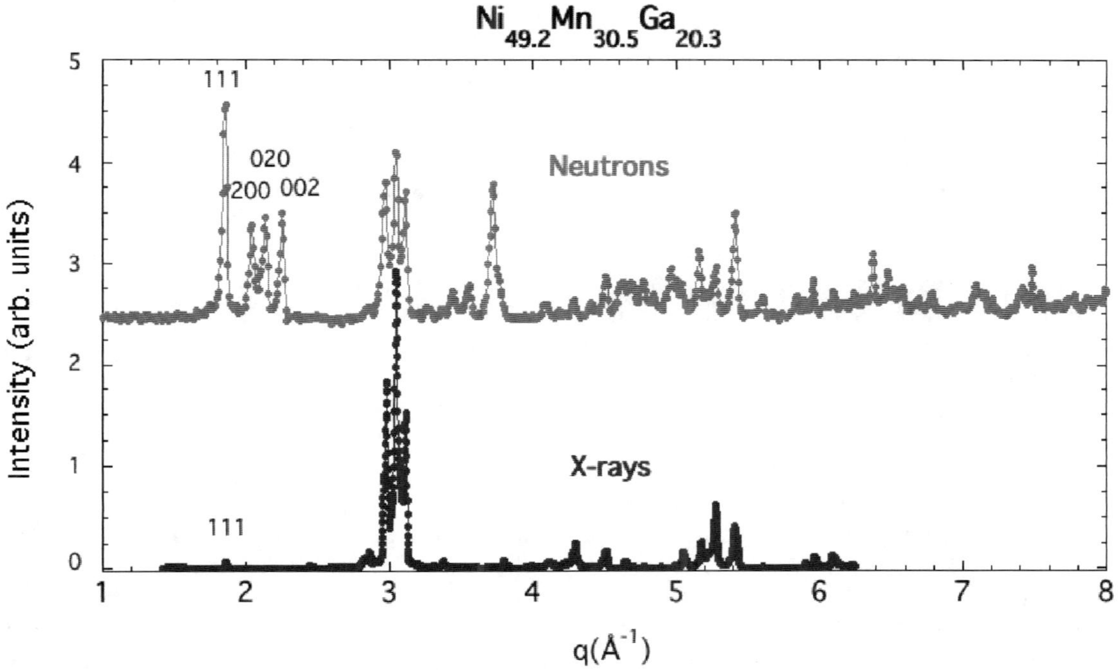

Figure 1. Comparison between X-rays and neutron diffraction patterns in a Magnetic Shape Memory Alloy. Note the intense superstructure peaks in the neutron diffraction are absent in the X-rays. Neutron intensity at large scattering angles are much more intense as well.

The synchrotron radiation is in fact complementary to neutrons for it is so strong that minute samples can be used with full resolution and large signal to noise ratio. Strong interaction with electrons provide data about the electronic structure too. Moreover, X-ray Magnetic Circular Dichroism (XMCD) provides atom selective magnetic information that complements the site selective information of neutrons.

In this paper we present a short survey of scattering experiments with neutrons, as well as X ray Photoemission Spectroscopy (XPS) and XMCD, which are complementary to elucidate the electronic and magnetic character of MSMA and the changes occurring at the martensitic transformation.

2. INS and XPS at the Martensitic transition

The Martensitic transformation has a number of possible final states including modulated structures extending over 5 or 7 lattice cells. In the prototype is the stoichiometric Ni_2MnGa, there is a pre-martensitic transformation that frozen the modulated structure in the cubic lattice, before the Martensitic transformation to a tetragonal L21 phase takes place. The "premartensitic" transformation from the cubic parent phase leads to a near-cubic modulated intermediate phase by means of of a soft phonon mode mechanism. A Zehludev, SM Shapiro et al. studied this transition using Inelastic Neutron Scattering (INS) in the 90's [12]. Recently V. Recarte, J. I. Pérez-Landazabal et al. [13] have shown the role of the magnetoelastic coupling in the transformation using inelastic neutron scattering under magnetic field. The experiment shows a temperature shift of the characteristic energy dip of the transverse acoustic branch TA2 under the magnetic field. Determination of the phonon dispersion relationship is a typical Neutron task and gives a deep insight into the mechanism of the martensitic transition and the modulation effects.

On the other hand, it has been proposed early in the study of MSMAs, that the main parameter driving the martensitic transformation and magnetic properties of FSMA is the electron concentration or average number of valence electrons for atom (e/a) so the electronic structure can be a key for the interpretation of the stability of some alloys that do not transform. The electronic structure of FSMAs, or at least their density of states (DOS), can be determined by X-ray Photoemission Spectroscopy (XPS), which is a technique available in many laboratories. Soft X rays, however, can only probe a few atomic distances from the surface of the material. To get a deeper probe, representative of the bulk, we need high energy X-rays and, therefore, a synchrotron source is essential.

In this way, XPS experiments at high energy (8keV) have been performed recently in Spring 8 onto two different NiMnGa samples with large differences in their electron concentration, so the low e/a alloy does not show the martensitic transition (MT) [14]. The following findings were obtained from the experiments:

- The spectral weight near EF is larger for larger e/a
- Ni 3d eg peak shifts to higher energy by 0.2 eV upon MT
- Rigid band model is not applicable for different e/a
- Jahn-Teller distortion scenario explain the MT al high e/a
- For low electron concentration, the electron energy gain in the split band is not sufficient to overcome the energy loss from the lattice distortion and no MT occurs.

3. Preferential site occupancy and magnetic moment

Electron concentration can be taken as the main parameter driving transformation and magnetic properties of FSMAs as a first approximation. However, in off stoichiometric alloys the magnetic moment is not a single function of e/a but the exact distribution of magnetic atoms in the lattice must be taken into account. For instance, the increase of order upon annealing in a rapidly quenched MSMA ribbon induces the corresponding increase of the magnetization (34%), martensitic transformation (>10K) and Curie temperature (>20K) [15], so that an alloy with different atomic order has different transformation and magnetic properties. The evolution of chemical order upon annealing in MSMAs has been studied by Recarte et al. [16] using neutron diffraction experiments. Another technique that allows changes in atomic order to be followed is EXAFS. This was applied to the rapidly quenched ribbon already mentioned with composition $Ni_{51}Mn_{28}Ga_{21}$ by Chaboy et al. [17]. Measurements were carried out at Spring 8 in Japan.

The radial distribution function (RDF) around Mn and Ni atoms are shown in fig.2 in the as quenched state and after annealing. This function is obtained as the Fourier transform of the EXAFS function obtained in the K absorption edge, and reflects the atomic distribution around a given atom. It is clear that the Ni atoms have its neighborhood unchanged by the annealing while the Mn atoms undergo a drastic change that correspond to an ordering process reflected in the higher amplitude and lower width of the RDF. This indicates that the ordering process leaves the Ni atoms unchanged but Mn ones reorganize and lower the randomness around them.

Journal of Physics: Conference Series **663** (2015) 012014 doi:10.1088/1742-6596/663/1/012014

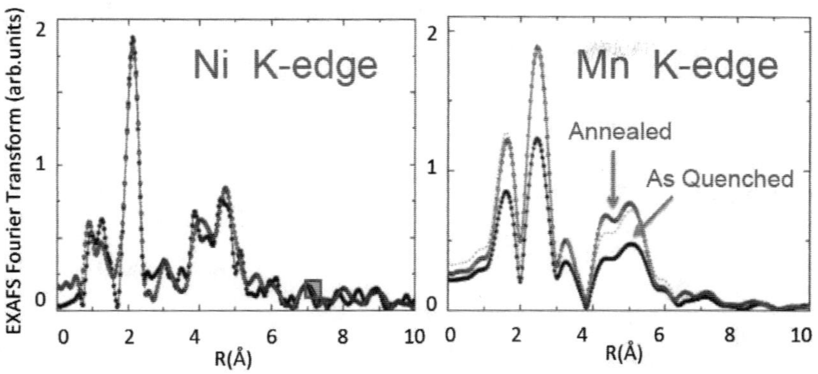

Figure 2. Radial distribution function around Ni and Mn atoms in a rapidly quenched ribbon of composition $Ni_{51}Mn_{28}Ga_{21}$, both as quenched and after 18 hours annealing at 315 °C. The Mn atoms undergo a clear reorganization but Ni ones remain unchanged.

Another interesting feature is the distribution of excess atoms among the vacant sites in off-stoichiometric alloys. In order to determine the preferential site occupancy as a function of the composition, neutron diffraction experiments have been carried out in a series of NiMnGa alloys. The experiments were performed at the high flux diffractometer D20 at the Institute Laue Langevin (ILL) in Grenoble, France. Main results are summarized in table 1. More details are given in [18,19].

Table 1. Site occupation of the atoms in the studied alloys, as deduced from the neutron diffraction experiments. The Wyckoff nomenclature is used together with the most simple one related to the atoms that occupy the site in the stoichiometric Ni2MnGa alloy.

Site (Wyckoff)	Ni44Mn38Ga18			Ni49Mn31Ga20			Ni50Mn29Ga21			Ni52Mn26Ga22		
	Ni	Mn	Ga	Ni	Mn	Ga	Ni	Mn	Ga	Ni	Mn	Ga
Ni (8c)	1.79	0.21	0.0	1.97	0.03	0.0	2.0	0.0	0.0	2.0	0.0	0.0
Mn (4a)	0.0	1.0	0.0	0.0	1.0	0.0	0.0	1.0	0.0	0.08	0.92	0.0
Ga (4b)	0.0	0.25	0.75	0.0	0.19	0.81	0.0	0.16	0.84	0.0	0.12	0.88

The following conclusions can be drawn: For alloys with Mn excess and defect of Ni and Ga, the Mn excess occupies the Ni and Ga vacant positions. In those alloys with both Ni and Mn excess, the Ni atoms go to Mn positions, displacing properly sited Mn atoms to Ga positions.

Now we can try to apply a localized magnetic moment model based on the early values of the moment obtained in the stoichiometric alloy Ni2MnGa by J Brown et al. [2] and Islam et al. [20]. The main assumptions of our model are:

a) The magnetic moment of the alloys is localized. Mn has 3.51μB and Ni 0.33μB
b) Mn atoms couple ferromagnetically (FM) when located at proper sites
c) Mn atoms at shorter distances have antiferromagnetic (AF) interactions
d) Ni always couple ferromagnetically to proper sited Mn
e) Coupling between off site Mn atoms is given by the Bethe Slater curve (Fig. 3)

The above model allows to calculate the average magnetic moment of any alloy, if the site occupancy is known. As depicted in fig. 3 we assume AF coupling between MnMn and MnGa atoms in excess Ni alloys. In defect Ni alloys, however, the Ni positions occupied by Mn atoms have the strongest AF coupling with MnMn ones, so that they can upturn the AF coupling between adjacent MnMn and MnGa atoms to FM. Taking into account the above considerations we obtain the magnetic moment values in table 2. They show a much better agreement with experimental values than the predictions of the rigid band model, also included for comparison.

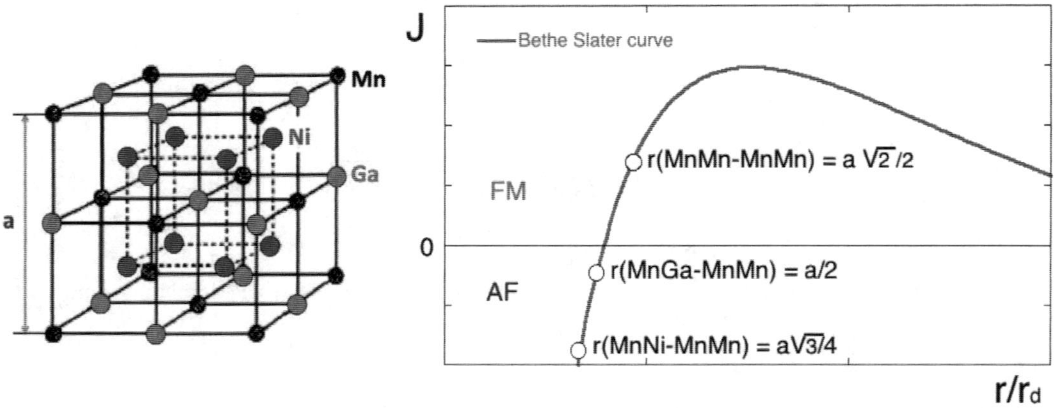

Figure 3. L21 crystal structure of Ni_2MnGa in the austenitic phase and exchange interactions between Mn atoms at different sites of the structure, assuming Bethe-Slater exchange variation with distance.

Table 2. Experimental and calculated moments of some studied alloys. All moments are given in Bohr magnetons (μ_B) per formula unit. e/a = electron concentration; μ_{EXP} = experimental moment determined at 5 K; μ_{RB} = calculated moment in the Rigid Band model; μ_{LOC} = calculated moment in the localized model.

Alloy	e/a	μ_{EXP}	μ_{RB} (error)	μ_{LOC} (error)
Ni44Mn38Ga18	7.58	3.06	2.37 (22.4%)	2.96 (3.3%)
Ni49Mn31Ga20	7.67	3.41	3.39 (0.6%)	3.46 (1.5%)
Ni50Mn29Ga21	7.67	3.60	3.60 (0.1%)	3.55 (1.2%)
Ni52Mn26Ga22	7.68	3.59	4.02 (12.0%)	3.48 (3.1%)

4. Direct determination of the magnetic moment

While the above model, together with the determination of site occupancy of the different magnetic atoms, gives quite good values for the total magnetic moment of the studied alloys, several points, as the possible contribution of conduction electrons to the magnetic moment, remain unclear and the most direct determination of the magnetic moment at each crystallographic site is welcome. For this to be achieved, polarized neutron scattering (PNS) in single crystals is a perfect tool.

PNS has been recently performed in single crystals of compositions $Ni_{52}Mn_{26}Ga_{22}$, $Ni_{49}Mn_{30}Ga_{21}$ and $Ni_{43}Mn_{38}Ga_{19}$, grown by the Bridgman method at the Ames Laboratory (Iowa, USA) [21]. The compositions are almost isoelectronic, but cover a quite large range of Ni concentrations as to be representative for testing the previous localized moment model. The nuclear structure of the alloys was determined at the ILL on D10 and D15 instruments. Polarized neutron diffraction measurements were carried out at D3 in the austenitic phase. Data analysis was performed using flipping ratios fits with the FullProf software. Maximum Entropy method [22] MemSys code, available at the ILL, was used for direct visualization of the atomic magnetization density maps using the Fourier transform of the diffraction data. A typical result is depicted in fig 4. It shows the density of magnetic moment along 111 plane in the L21 Austenitic phase of the $Ni_{53}Mn_{27}Ga_{20}$ alloy. The data were taken at 340 K under a magnetic field of 2 Tesla, which largely saturates the sample.

The results confirm the previous model in different aspects:

a) Magnetic moments are almost fully localized at Mn atoms in Mn positions (MnMn). Ni positions also show a parallel moment but much lower, as previously assumed. Ga positions show antiparallel magnetic moments arising from the excess Mn atoms located at such positions, coupling antiferromagnetically to the MnMn ones.

b) The average magnetic moment value at Ni positions (NiNi) for the Ni defective alloy is lower than in the other ones. This is expected from the presence of excess Mn atoms at Ni positions, coupling antiferromagnetically with the total moment.

c) MnMn atoms show small average moment values in the excess Ni composition, thus confirming that the excess Ni occupy the Mn positions and decrease the total magnetic moment at the Mn positions, due the lower magnetic moment of the Ni atoms.

d) The average moment of MnGa atoms is smaller in the lowest Ni content alloy. This supports our model of both ferromagnetic and antiferromagnetic coupling of the MnGa atoms depending on the neighbor Ni sites being occupied by Mn atoms or not.

The above findings strongly support the localized model for the magnetism in NiMnGa, and the scheme of FM and AF interactions among Mn atoms as a function of the atomic distance. On the other hand, the magnetic moment per Mn atom deduced from the experiment is different for each alloy, despite the reduced measuring temperature of all of them being the same ($T_{meas}/T_{Curie} = 0.92$) [22]. This finding can be of great help for first principles calculations of the magnetic moment in such alloys.

As shown, direct determination of the magnetic moment on the crystal positions by PND is a formidable tool to know the magnetic structure of the off stoichiometric NiMnGa FSMAs. However, the complementary knowledge of the site occupation by the different atoms is a previous requirement to interpret the results.

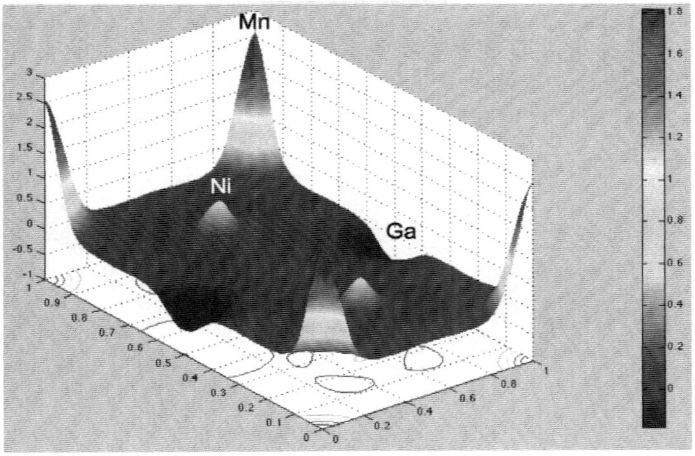

Figure 4. Magnetic moment distribution of Ni53Mn27Ga20 along the 111 plane of the L21 structure as to include all atomic positions. Labeling refers to stoichiometric positions of the corresponding atoms. The magnetic moment is mainly localized at the Mn positions. A much lower value appears at Ni positions, as expected. The negative moment at Ga sites is due to the excess Mn atoms located in such positions.

In the case of FSMAs with more than two magnetic atoms per unit cell, the above procedure is useless, because there are quite a number of possible magnetic moments assigned to each atom that can give results close to each other. A different technique, X rays Magnetic Circular Dichroism (XMCD), gives the magnetic moment of each specific atom, whatever the position in the crystal lattice it occupies, and is more desirable for such compounds.

Recently the determination of the magnetic moment by XMCD has been carried out in a three magnetic component ferromagnetic shape memory alloy (FSMA): $Ni_{49}Fe_{17.5}Co_{4.5}Ga_{29}$ [23]. The sample was prepared in the form of a thin film by sputtering, in order to maximize the volume explored by the beam using a low incident angle. Atomic moments have been determined at room temperature (Austenite) and at 50K (Martensite), and the results are depicted in fig. 5. The main conclusion is that the three atoms couple FM in both phases and the average moment of each atomic species among all the crystal positions can be obtained independently. This conclusion is really specific of XMCD and couldn't be obtained by neutrons. Both techniques are, therefore, complementary.

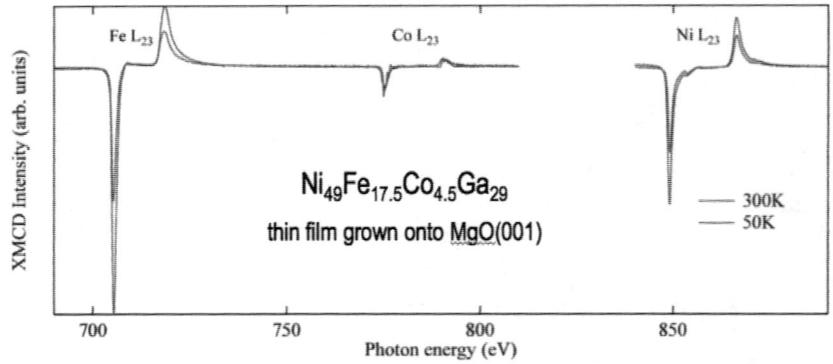

Figure 5. Magnetic Circular Dichroism in a FSMA thin film with three different magnetic atoms in its composition.

A different kind of magnetic shape memory alloys from the NiMnGa discussed before, are those showing Meta-Magnetic behavior (MMSMA). Their composition includes In, Sn or Sb instead of Ga (see, e.g., [24] and references therein). In such alloys the martensite, just below the transformation temperature, has a vanishing saturation magnetization (fig. 6), because of this, the reverse transformation into austenite can be induced by a moderate magnetic field [24-26]. The magnetic character of the martensitic phase is controversial and several options have been proposed, such as spin glass [27, 28], antiferromagnetic [29] or simply paramagnetic [30].

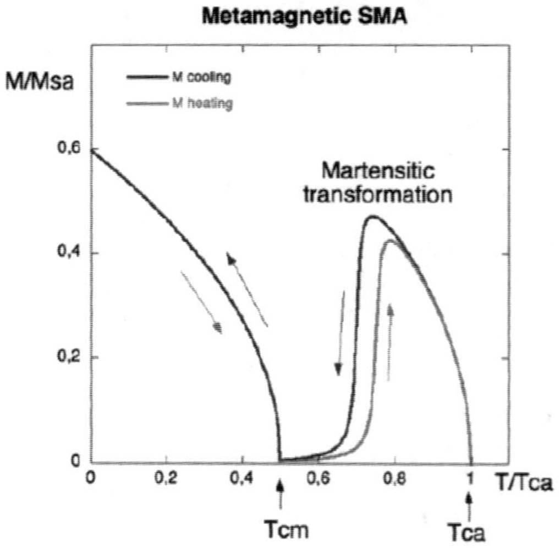

Figure 6. Schematic evolution of the saturation magnetization in a MetaMagnetic Shape Memory Alloy (MMSMA) showing a "non magnetic" state below the martensitic transformation temperature and above the martensitic Curie temperature T_{cm}.

Journal of Physics: Conference Series **663** (2015) 012014 doi:10.1088/1742-6596/663/1/012014

A recent study by XMCD [31] on a MMSMA thin film of composition Ni48Co5Mn35In12, has found an antiparallel alignment of Mn and Ni/Co magnetic moments in both Austenite and Martensite phases. Theoretical first principles calculations agree with the finding. The XMCD reveals itself as a powerful technique that will see an increasing demand in the next years.

5. Conclusions

Both neutrons and synchrotron radiation have a great role to play in the study of Magnetic Shape Memory alloys. Inelastic Neutron Scattering is the perfect tool for phonon studies at the structural transformations in SMA and FSMAs, while XPS have a complementary role in determining the electronic mechanisms of the transformation, that has been attributed to a Jahn-Teller distortion. Neutron diffraction is much more powerful than X-rays for superstructure studies and site occupancy determination in NiMnGa alloys, both as a function of composition and thermal treatments, because of the very different scattering lengths for the three atoms. XANES ands EXAFS are complementary to Neutron diffraction to follow the chemical order evolution upon annealing.

Polarized neutron diffraction (PND) experiments allow to determine the localized moment of Ni and Mn atoms in off stoichiometric NiMnGa single crystals, provided site occupancy is simultaneously known. XMCD is complementary to PND for determining the magnetic moment as it is atom specific. In case of more that two magnetic atoms in the structure XMCD is unique in giving a detailed picture of the magnetic coupling.

Acknowledgments

The authors want to thank B. Ouladiaff, C. Mondeli, T. Hansen and A. Stalnaut for assistance in the neutron diffraction measurements at ILL, and J. Chaboy, C. A. Jenkins, N. Kawamura and K. Sumida for XANES and XMCD experiments in Spring 8.

Funding by the Spanish Ministry Economy and Competitiveness (grant MAT2014-56116), the Basque Departments of Industry (ACTIMAT project) and Education (funding of consolidated research groups of the University of the Basque Country) is greatly acknowledged.

References

[1] Webster P J, Ziebeck K R A, Town S L and Peak M S 1984 *Philos. Mag. B* **49** 295
[2] Brown P J, Bargawi A Y, Crangle J, Neumann K-U and Ziebeck K R A 1999 *J. Phys.: Condens. Matter.* **11** 4715
[3] Pons J, Chernenko V A, Santamarta R and Cesari E 2000 *Acta Mater.* **48** 3027
[4] Enkovaara J, Ayuela A, Jalkanen J, NordstroÅNm L and Nieminen R M 2003 *Phys. Rev. B* **67** 054417
[5] Barman S R, Banik S and Chakrabarti A 2005 *Phys. Rev. B* **72** 184410
[6] Chakrabarti A, Biswas C, Banik S, Dhaka R S, Shukla A K and Barman S R 2005 *Phys. Rev. B* **72** 073103
[7] Brown P J, Crangle J, Kanomata T, Matsumoto M, Neumann K-U, Ouladdiaf B and Ziebeck K R A 2002 *J. Phys.: Condens. Matter* **14** 10159
[8] Brown P J, Gandy A P, Kanomata T, Matsumoto M, Neummann K, Neumann K-U, Sheikh A and Ziebeck K R A 2008 *Mater. Sci. Forum* **583** 285
[9] Barandiaran J M, Gutiérrez J, Lázpita P and Feuchtwanger J 2011 *Mater. Sci. Forum* **684** 73
[10] Bacon G E 19777 *Neutron scattering in chemistry* (Butterworths, London and Boston)
[11] 2003 *Neutron data booklet* (Eds. Dianoux A J and Lander G, Institute Laue-Langevin)
[12] Zheludev A, Shapiro S M and Wochner P 1996 *Phys. Rev. B* **54** 15045
[13] Recarte V, Perez-Landazabal J I, Sanchez-Alarcos V, Cesari E, Jimenez-Ruiz M, Schmalz K and Chernenko V A 2013 *App. Phys. Lett.* **102** 201906

[14] Kimura A, Ye M, Taniguchi M, Ikenaga E, Barandiaran J M and Chernenko V A 2013 *Appl. Physics. Lett.* **103** 072403
[15] Gutiérrez J, Barandiarán J M, Lázpita P, Seguí C and Cesari E 2006 *Sensors and Actuators A* **129** 163
[16] Recarte V, Perez-Landazabal J I, Gomez-Polo C, Sanchez-Alarcos V, Cesari E and Pons J 2010 *J. Phys. Condens. Matter* **22** 416001
[17] Chaboy J, Lázpita P, Barandiarán J M, Gutiérrez J, Fernández-Gubieda M L and Kawamura N 2009 *J. Phys.: Condens. Matter.* **21** 016002
[18] Richard M L, Feuchtwanger J, Allen S M, O'Handley R C, Lazpita P, Barandiaran J M, Gutierrez J, Ouladdiaf B, Mondelli C, Lograsso T and Schlagel S 2007 *Phil. Mag.* **87** 3437
[19] Lázpita P, Barandiarán J M, Gutiérrez J, Feuchtwanger J, Chernenko V A and Richard M L 2011 *New Journal of Physics* **13** 033039
[20] Islam A, Haskel D, Lang J C, Srajer G, Lee Y, Harmon B, Goldman A I, Schlagel D L and Lograsso T A 2006 *J. Magn. Magn. Mater.* **303** 20
[21] Lázpita P, Gutierrez J, Barandiaran J M, Chernenko V A, Mondelli C and Chapon L 2014 *J. Phy,: Conf. Series* **549** 012016
[22] Papoular R J and Gillon B 1990 *Europhys. Lett.* **13** (5) 429
[23] Sumida K, Shirai K, Zhu S, Taniguchi M, Ye M, Ueda S, Takeda Y, Saitoh Y, Rodriguez I, Barandiaran J M, Chernenko V A and Kimura A (*to be published*)
[24] Planes A, Manosa Ll and Acet M 2009 *J. Phys.: Cond. Mat.* **21** 233201
[25] Kainuma R, Imano Y, Ito W, Sutou Y, Morito H, Okamoto S, Kitakami O, Oikawa K, Fujita A, Kanomata T and Ishida K 2006 *Nature* **439** 957
[26] Sharma V K, Chattopadhyay M K, Chouhan A and Roy S B 2009 *J. Phys. D: Appl. Phys.* **42** 185005
[27] Cong D Y, Roth S, Liu J, Luo Q, Pötschke M, Hürrich C and Schultz L 2010 *Appl. Phys. Lett.* **96** 112504
[28] Pramanick S, Chattopadhyay S, Giri S, MajumdarS and Chatterjee S 2014 *J. Appl. Phys.* **116** 083910
[29] Liao P, Jing C, Wang X L, Yang Y J, Zheng D, Li Z, Kang B J, Deng D M, Cao S X, Zhang J C and Lu B 2014 *Appl. Phys. Lett.* **104** 092410
[30] Umetsu R Y, Kainuma R, Amako Y, Taniguchi Y, Kanomata T, Fukushima K, Fujita A, Oikawa K and Ishida K 2008 *Appl. Phys. Lett.* **93** 042509
[31] Klaer P, Herper H C, Entel P, Niemann R, Schultz L, Fähler S and Elmers H J 2013 *Phys. Rev. B* **88** 174414

Institute of Physics
Dirac House, Temple Back
Bristol BS1 6BE UK

ISSN: 1742-6588
ISBN 978-1-5108-1735-7

CURRAN ASSOCIATES INC.
proceedings.com